LEIBNIZ'S METAPHYSICS OF NATURE

THE UNIVERSITY OF WESTERN ONTARIO SERIES IN PHILOSOPHY OF SCIENCE

A SERIES OF BOOKS

IN PHILOSOPHY OF SCIENCE, METHODOLOGY,

EPISTEMOLOGY, LOGIC, HISTORY OF SCIENCE,

AND RELATED FIELDS

VOLUME 18

NICHOLAS RESCHER

Department of Philosophy,
University of Pittsburgh

LEIBNIZ'S METAPHYSICS OF NATURE

A Group of Essays

D. REIDEL PUBLISHING COMPANY

DORDRECHT : HOLLAND / BOSTON : U.S.A.

LONDON : ENGLAND

Library of Congress Cataloging in Publication Data

Rescher, Nicholas.
Leibniz's metaphysics of nature.

(The University of Western Ontario series in philosophy of science; v. 18)
Includes bibliographical references and indexes.
1. Leibniz, Gottfried Wilhelm, Freiherr von, 1646–1716–Metaphysics.
2. Metaphysics. 3. Philosophy of nature. 4. Space and time. I. Title.
II. Series: University of Western Ontario series in philosophy of science; v. 18.
B2599.M7R47 110′.92′4 81-5944
ISBN 90-277-1252-2 AACR2
ISBN 90-277-1253-0 (pbk.)

Published by D. Reidel Publishing Company,
P.O. Box 17, 3300 AA Dordrecht, Holland.

Sold and distributed in the U.S.A. and Canada
by Kluwer Boston Inc.,
190 Old Derby Street, Hingham, MA 02043, U.S.A.

In all other countries, sold and distributed
by Kluwer Academic Publishers Group,
P.O. Box 322, 3300 AH Dordrecht, Holland.

D. Reidel Publishing Company is a member of the Kluwer Group.

Printed in The Netherlands

For Yvon Belaval

CONTENTS

PREFACE

The essays included in this volume are a mixture of old and new. Three of them make their first appearance in print on this occasion (Nos III, IV, and V). The remaining four are based upon materials previously published in learned journals or anthologies. (However, these previously published papers have been revised and, generally, expanded for inclusion here.) Detailed acknowledgement of prior publications is made in the notes to the relevant articles. I am grateful to the editors of these several publications for their kind permission to use this material.

I am grateful to an anonymous reader for the Western Ontario Series for some useful corrigenda. And I should like to thank John Horty and Lily Knezevich for their help in seeing this material through the press.

NICHOLAS RESCHER
Pittsburgh, Pennsylvania
May, 1980

INTRODUCTION

The unifying theme of these essays is their concern with Leibniz's metaphysics of nature. In particular, they revolve about his cosmology of creation and his conception of the real world as one among infinitely many equipossible alternatives.

The essays address themselves to the following issues: the theory of creation as a selection among infinitely diversified alternatives, the cognitive problems that confront man's finite intellect in scientific inquiry at a world of infinite complexity and diversity, the central role of systematization in human cognition, the characteristic nature of contingent truth as a matter of infinitely ramified comparisons, the infinite ramifications of the relations among the substances comprising one self-contained world-system, the infinite diversity of alternative space-time frameworks for different possible worlds, and the contributions of Leibniz's work on the infinitesimal calculus to his creation-metaphysics.

As is clear, even from this barebones summary, the idea of infinity runs through Leibniz's philosophy of nature as an ever-recurrent theme. He was the true child of a Renaissance culture for which the finitude of the classical world had burst asunder once and for all. Nevertheless, he continued true to his Greek heritage in insisting on the systemalizability of any and all rational knowledge of the real. Philosophically it was not for nothing that Leibniz was the inventor of the infintesimal calculus. For it was this that provided him with the idea that one could systematize with mathematical rigor thought regarding matters of infinite complexity and diversity. The calculus was Leibniz's model for continuing the classical architectonic of antiquity with the infinitistic richness of the post-Renaissance thought-world of modern science.

A brief word on bibliographical matters. Leibniz's philosophical writings are here frequently cited with reference to C. I. Gerhardt, *Die philosophischen Schriften von G. W. Leibniz*, seven volumes, Weidmann, Berlin (1875–90). This is cited as *GP*.

For English translations of Leibniz's writings I have generally relied on Leroy E. Loemker, *Gottfried Wilhelm Leibniz; Philosophical Papers and Letters*, 2nd edn. D. Reidel, Dordrecht and Boston (1969). This is cited as *Loemker*.

For further and fuller bibliographical information the reader may refer to Loemker's book or to the present writer's *Leibniz: An Introduction to his Philosophy*, Basil Blackwell, Oxford (1979).

ESSAY I

LEIBNIZ ON CREATION AND THE EVALUATION OF POSSIBLE WORLDS

1. STAGESETTING

Present-day studies of Leibniz have remarkably little to say about one altogether central doctrine of his metaphysical system – his theory of the *standard of value* determinative of the relative goodness of possible worlds. Leibniz's manifold discussions of the metaphysical perfection of worlds are given short shrift, presumably being viewed by most recent commentators as something of a vestigial remnant of scholastic sophistry. This is unfortunate for his teachings on this topic are, in fact, among the most interesting and significant aspects of Leibniz's entire system.

Which of the alternative possible worlds he contemplates *sub ratione possibilitatis* is God to select for actualization? This question clearly poses one of the central issues of Leibniz's philosophy. He bitterly opposed the position of Descartes and Spinoza, whom Leibniz took to maintain the indifference and arbitrariness of God's will. He emphatically maintains that:

> [W]hen we say that things are not good by any rule of excellence but solely by the will of God, we unknowingly destroy, I think, all the love of God and all his glory. For why praise him for what he has done if he would be equally praiseworthy in doing exactly the opposite? Where will his justice and wisdom be found if nothing is left but a certain despotic power, if will takes the place of reason, and if, according to the definition of tyrants, that which is pleasing to the most powerful is by that very fact just? Besides it seems that every act of will implies some reason for willing and that this reason naturally precedes the act of will itself. This is why I find entirely strange, also, the expression of certain other philosophers who say that the eternal truths of metaphysics and geometry, and consequently also the rules of goodness, justice, and perfection, are merely the effects of the will of God.[1]

The existence of an objective criterion of goodness for possible worlds wholly independent of the will of God is a crucial feature of The Principle of Perfection.[2] This principle is Leibniz's philosophic formulation of the theological principle of God's goodness. It asserts that in the creation of the world, God acted in the best way possible. Leibniz was eager to combat any manifestation of

the contrary opinion [which] seems to me extremely dangerous and to come very near to that of the latest innovators whose opinion it is that the beauty of the universe and the goodness which we ascribe to the works of God are nothing but the chimeras of men who think of him in terms of themselves. Then, too, when we say that things are not good by any rule of excellence but solely by the will of God, we unknowingly destroy, I think, all the love of God and all his glory.[3]

Leibniz thus held that possibilities are objectively good or bad by a standing "*règle de bonté*" which operates altogether independently of the nature of existence and of the will of God. Indeed this standard of goodness is the basis for two crucial modal distinctions: (1) since God acts for the best in all of his actions, that of creation preeminently included, it serves to demarcate the actual from the possible, and (2) given God's goodness, it renders the sphere of the actual as *necessary* – not in the absolute or metaphysical sense of this term, but in its relative or moral sense. The standard of goodness is accordingly pivotal for the operation of these modal distinctions in the system of Leibniz.

Leibniz repeatedly says that all possible substances have "a certain urge (*exigentia*) toward existence"; since to be a possible is to be a possibly *existing* thing, we find in possibles a *conatus*, a dynamic striving toward existence proportionate to the perfection of the substance. This way of talking has led certain commentators, primarily A. O. Lovejoy,[4] to argue that Leibniz's talk of the selection of a (i.e., the best) possible world for actualization by God is a pointless redundancy, since it is the intrinsic nature of these substances to prevail in the struggle for existence among alternative possibilities. But this approach misconceives the issue badly, for it is only because God – in view of his *moral* perfection – *has chosen* to subscribe to a certain standard of *metaphysical* perfection in selecting a possible world for actualization that possible substances come to have (figurative) "claim" to existence. God cannot be bypassed here. Exigency to existence is a matter of exerting a certain attraction in the setting of God's deliberations regarding creation. The relationship between "quantity of essence" or "perfection" of substance on the one hand and, on the other, their claim or exigency to existence is *not* a logical linkage at all – a thesis which would reduce Leibniz's system into a Spinozistic necessitarianism – but a connection mediated by a free act of will on the part of God.

Leibniz again and again insisted that there is an independent standard of the perfection of things – a standard determined by considerations of objective necessity, which the preferences and decisions of the diety could alter no more than the sum of two plus two. Moreover, this standard is not to apply simply to the actual domain of the real world, but is operative throughout the modally variant sphere of the morely possible as well.

But just what is this criterion which a God who seeks to actualize *the best* of possible worlds employs in identifying it? By what criterion of merits does God determine whether one possible world is more or less perfect than another? This standard, Leibniz maintains, is the combination of *variety* and *order*. Accordingly, the best of possible worlds is that which successfully manages to achieve the greatest richness of phenomena (*richesse des effects (Discours de métaphysique,* §5), *fecondité* (*Théodicée,* §208) *varietas formarum* (*Phil. Werke*, ed. Gerhardt, VII, 303)) that can be combined with (*est en balance avec* (*Discours de métaphysique* §5), *sont les plus fécondes par rapport à* (*Théodicée,* §208)) the greatest simplicity of means (*la simplicité des voyes* (*Discours de métaphysique,* §5 and *Théodicée,* §208), *le plus grand ordre* (*Principles of Nature and of Grace*, § 10)).

In the elegant 1697 essay *De rerum originatione radicali,*[5] Leibniz puts the matter as follows:

Hence it is very clearly understood that out of the infinite combinations and series of possible things, one exists through which the greatest amount of essence or possibility is brought into existence. There is always a principle of determination in nature which must be sought by maxima and minima; namely, that a maximum effect should be achieved with a minimum outlay, so to speak. And at this point time and place, or in a word, the receptivity or capacity to the world, can be taken for the outlay, or the terrain on which a building is to be erected as commodiously as possible, the variety of forms corresponding to the spaciousness of the building and the number and elegance of its chambers.[6]

Of course, God could have chosen a world whose law structure would be nomically simpler; a world composed of one element, say iron, alone would not need the whole complex apparatus of biological laws. Again, the world could be richer in phenomena; it might include all manner of strange things and creatures. It is the distinguishing feature of this, the actual, and thus the best possible world that it manages to strike the optimal balance here. And it is just exactly this feature of optimal balance that recommends the world to a God who contemplates it as simply one alternative among others when confronting such alternatives *sub ratione possibilitatis* in the course of making a creation-choice.

It is worthwhile to consider in some detail the individual components of Leibniz's two-factor criterion of variety and richness of phenomena on the one hand and lawfulness or order on the other.

The reference to lawfulness clearly leads back straightaway to Greek ideas (balance, harmony, proportion). The origination of cosmic order is a key theme in the Presocratics (especially the Milesians) and becomes one of the great central issues of ancient philosophy with the writing of Plato's *Timaeus*. Afterwards, it of course plays a highly prominent role in the church fathers and achieves a central place in scholasticism in connection with the cosmological argument for the existence of God.

The prime factor in Leibniz's theory is not, however, lawfulness as such, but the simplicity or economy of laws. Leibniz, as we know, held that *every* possible world is lawful. As he puts it in §6 of the *Discours de mètaphysique: on peut dire que, le quelque manière que Dieu auroit crèè le monde; il auroit toujours estè regulier et dans un certain ordre general*. The critical difference between possible worlds in point of lawfulness is, thus, not whether there are laws or not – there *always* are – but whether these laws are relatively simple.

Regrettably, Leibniz nowhere treats in detail the range of issues involved in determining the relative simplicity of bodies of laws, and indeed he does not seem to be fully aware of the complexities that inhere in the concept of simplicity. Perhaps he did not think it necessary to go into details because it is, after all, by God and not us imperfect humans that this determination is to be made. And in any case, it is clear enough in general terms what he had in mind. No one for whom the development of classical physics in its "Newtonian" formulation as a replacement of Ptolemaic epicycles and Copernican complexity was a living memory could fail to have some understanding of the issues. (It might be observed parenthetically that Leibniz would surely have viewed with approval and encouragement the efforts by Nelson Goodman, J. G. Kemeny, and others during the 1950s[7] to develop an exact analysis of the concept of simplicity operative in the context of scientific theories.) So much then for simplicity of laws; let us turn to variety.

The situation as regards *variety* is even somewhat more complicated. As Leibniz considers it, variety has two principal aspects: fullness or completeness or comprehensiveness of content on the one hand, and diversity and richness and variation and complexity upon the other. All these factors are certainly found in ancient writers. They are notable in the *Timaeus*[8] and play a significant role in Plotinus and neoplatonism.[9] The church fathers also stressed the role of completeness and fullness (*fecunditas*) as a perfection, and it is prominent in St Thomas's treatment of the cosmological argument and also in the later schoolmen. The recognition of the metaphysical importance of variety is thus an ancient and stable aspect of Platonic tradition.

But a new, Renaissance element is present in Leibniz's treatment of this theme: the aspect of *infinitude* that did not altogether appeal to the sense of tidiness of the more fastidious Greco-Roman mentality. The Renaissance evolution of a spatially infinite universe from the finite cosmos of Aristotle is a thoroughly familiar theme. And it represents a development that evoked strong reactions. One may recall Giordano Bruno's almost demonic delight with the break-up of the closed Aristotelian world into one opening into an infinite universe spread throughout endless space. Others were not delighted but appalled – as, e.g., Pascal was frightened by "the eternal silence of infinite spaces" of which he speaks so movingly in the *Pensées* (§§ 205, 206). With Leibniz, on the other hand, the infinite always carries positive connotations.

An analogous development occurred with respect to the strictly *qualitative* aspects of the universe. Enterprising and ardent spirits like Paracelsus, Helmont, and Bacon delighted in stressing a degree of complexity and diversity not envisaged in the ancient authorities. A vivid illustration of this welcoming of diversity is Leibniz's insistence that the variety of the world is not just a matter of the number of its substances, but of the infinite multiplicity of the forms or kinds they exemplify. He would not countenance a *vacuum formarum*: but taught that infinite gradations of kind connect any two natural species. Accordingly, he was positively enthusiastic – as no classically fastidious thinker could have been – about the discovery by the early microscopists of a vast multitude of little squirming things in nature. Leibniz's concern for qualitative infinitude as an aspect of variety represents a distinctly modern variation of an ancient theme. In giving not only positive but even paramount value to variety, complexity, richness and comprehensiveness, Leibniz expresses, with characteristic genius, the Faustian outlook of modern European man.

His concern for an objective standard of cosmic valuation gives Leibniz a central place within the tradition of *evaluative metaphysics*. The paternity of this branch of philosophy may unhesitatingly be laid at Plato's door, but as a conscious and deliberate philosophical method it can be ascribed to Aristotle, whose preoccupation in the *Metaphysics*[10] with the ranking schematism of prior/posterior is indicative of his far-reaching concern with the evaluative dimension of metaphysical inquiry.[11]

Now as has been indicated, there is nothing new and original in a stress upon variety and order as aspects of the perfection of creation. Both of these factors are prominent in the philosophical atmosphere of the late seventeenth century. Thus, for example, in §17 of Book I of Malebranche's *Traitè de la Nature et de la Grace* (published in 1680) we read:

> Or ces . . . lois sont si simples, si naturelles, & en même temps si fécondes que quand on n'auroit point d'autres raisons pour juger que ce sont establies par celui qui agit toujours par les voies les plus simples, dans l'action duquel il n'y a rien qui ne soit réglé, & qui la proportionne si sagement avec son ouvrage, qu'il opere une infinité de merveilles par un très-petit nombre de voluntez.[12]

What can Leibniz add to this very Leibnizian passage?

What he adds is exactly to establish these two long-prominent *aspects* of the world's perfection as jointly operative and mutually conditioning criteria joined within *a single two-factor standard of the perfection* of a possible world. What is specifically characteristic of Leibniz is the idea of combination and balance of these factors in a state of mutual tension.

But why should it be plausible to take this step and establish variety and order as conjoint but potentially conflicting yardsticks of perfection? The basis of plausibility of this Leibnizian standard rests upon a whole network of analogies, three of which are clearly primary:

(1) *Art*. Throughout the fine arts an excellent production requires that a variety of effects be combined within a structural unity of workmanship. Think here of the paradigm of Baroque music and architecture.

(2) *Statecraft*. Excellence can only be achieved in the political organization of affairs when variety (freedom) is duly combined with lawfulness (=order and the rule of law).

(3) *Science*. Any really adequate mechanism of scientific explanation must succeed in combining a wide variety of phenomena (fall of apple, tides, moon) within the unifying range of a simple structure of laws (gravitation).

All of these diverse paradigms meet and run together in Leibniz's thinking. Like lesser luminaries such as Rudolf Spengler and Ernst Cassirer, Leibniz had an extraordinarily keen eye for the perception of deep subsurface analogies. A true systematizer, he likes to exploit vast overarching connections and exhibits an extraordinary talent for transmitting a discernment of common structures into the formulation of an illuminating and fruitful theory.

2. MATHEMATICO-PHYSICAL INSPIRATION

The criterion of goodness or "perfection" for possible worlds is set forth by Leibniz in the following terms:

> God has chosen [to create] that world which is the most perfect, that is to say, which is at the same time the simplest in its hypotheses [i.e., its laws] and the richest in phenomena.[13]

The characteristic properties of each substance change from one juncture to another in accordance with its program. The properties of substance #1 at one juncture may be more or less in accordance with and thus reflected or mirrored in those of substance #2 at this juncture. Out of these mirroring relationships grow the regularities which represent the "hypotheses", the natural laws of the possible worlds. The "best", most perfect possible world is that which exhibits the greatest *variety of its contents* (richness of phenomena) consonant with the greatest *simplicity of its laws*:

> The ways of God are those most simple and uniform . . . (being) the most productive in relation to the *simplicity of ways and means*. It is as if one said that a certain house was the best that could have been constructed at a given cost . . . If the effect were assumed to be greater, but the process less simple, I think one might say, when all is said and done, that the effect itself would be less great, taking into account not only the final effect but also the mediate effect. For the wisest mind so acts as far as is possible, that the means are also ends of a sort, i.e., are desirable not only on account of what they do, but on account of what they are.[14]

We may think of the possible worlds as positioned along a curve of feasible order/variety combinations of the following general sort:

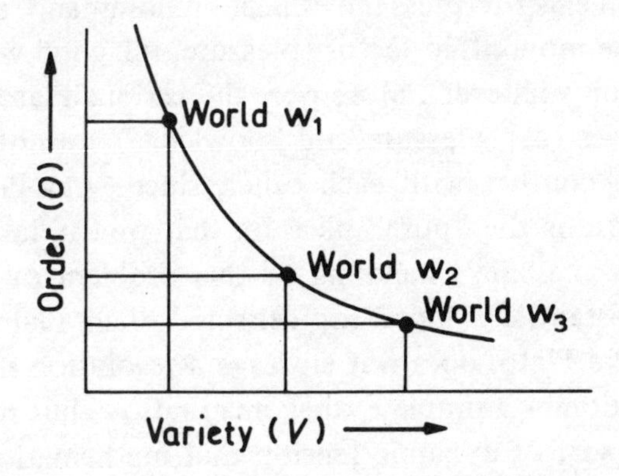

World w_1 is very orderly but lacks variety (it may, for example, be a vast sphere of copper suspended in air). World w_3 may be highly variegated – a virtually chaotic mixture of things – but lacking in order. World w_2 affords a much better mix. Here the order/variety combination value is as large as can be realized within the realm of realizability.

One immediately striking feature of the Leibnizian standard of metaphysical perfection in terms of orderliness and variety is that this is a *conflict-admitting two-factor criterion* and, as such, contrasts sharply with the long series of monolithic *summum bonum* theories that have so generally been in vogue in ethics – both before the time of Leibniz and afterwards, down to our own day.

From this aspect, Leibniz's position is strikingly reminiscent of that taken by his philosophical idol, Plato, in the *Philebus*. Plato there holds, as against various simplistic philosophical doctrines, that the good life cannot rest on (say) knowledge or beauty or pleasure alone, but requires a *mixture* of such factors in which each constituent has its proper share. But, of course, most of the subsequent ethical tradition from the Stoics and Epicureans to the utilitarians and Freudians were impatient with Platonic complexity and have been eager to press the ethical predominance and primacy of one unique monolithic factor (pleasure, the good will, personal adjustment, or whatever). Moreover, the various Platonic ingredients themselves (say pleasure and knowledge) are not in any obvious way in conflict with each other, since – as Plato himself stresses – there is the "pure" pleasure that we can take in knowledge as such. Leibniz's handling of this problem of mixture is, however, substantially more sophisticated than Plato's, because Leibniz, unlike Plato, does not envisage a resolution in terms of a fixed proportion – a simple Pythagorean ratio – but rather one in terms of the sort of dynamic tension that mathematical interrelationships of a more complicated (nonlinear) type make possible.

Put in economic terminology, Leibniz thinks of these factors as related, not by a fixed exchange ratio, but by variable trade-offs with diminishing marginal returns for both parameters.

The immediately striking feature of the criterion is that the two operative factors are *opposed* to one another and pull in opposite directions. On the one hand, a world whose only metal is (say) copper, or whose only form of animal life is the amoeba, will obviously have a simpler structure of laws because of this impoverishment. On the other hand, a world whose laws are more complex than the rules of the astrologers demands a wider variety of occurrences for their exemplification. Clearly the less variety a world contains – the more monotonous and homogeneous it is – the simpler its laws will be; and the more complex its laws, the greater the variety of its phenomena must be to realize them. Too simple laws produce monotony; too varied phenomena produce chaos. So these two criterial factors of order and variety are by no means cooperative, but stand in a relationship of mutual tension and potential opposition.

Taken together, variety and order provide a measure of the quantity of (potential) being or existence in reality that God seeks to maximize in his creation choice:

> We must also say that God makes the maximum of things he can, and what obliges him to seek simple laws is precisely the necessity to find place for as many things as can be put together: if he made use of of other laws, it would be like trying to make a building with round stones, which make us lose more space than they occupy.[15]

But it would be misleading to think of the maximization process as addressing itself to a single quantity ("quantity of essence") since this is itself a function of several distinct parameters (specifically including variety and order). It is just exactly the internal complexity and diversity at issue in "quantity of being" that is the object of characteristically creative Leibnizian insight.

Leibniz's conception of the deity's way of proceeding in the selection of one of the possible worlds for actualization can be represented and illustrated by the sort of infinite-comparison process familiar from the calculus and the calculus of variations. In his interesting and very important little essay *Tentamen Anagogicum* (ca. 1696), Leibniz puts the matter as follows:

> [T] he principles of mechanics themselves cannot be explained geometrically, since they depend on more sublime principles which show the wisdom of the Author in the order and perfection of his work. The most beautiful thing about this view seems to me to be that the principle of perfection is not limited to the general but descends also to the particulars of things and of phenomena and that in this respect it closely resembles the method of *optimal forms*, that is to say, of forms *which provide a maximum or minimum*, as the case may be – a method which I have introduced into geometry in addition to the ancient method of *maximal and minimal quantities*. For in these forms or figures the *optimum* is found not only in the whole but also in each part, and it would not even suffice in the whole without this. For example, if in the case of the curve of shortest descent between two given points, we choose any two points on this curve at will, the part of the line intercepted between them is also necessarily the line of shortest descent with regard to them.[16]

In taking as measure of perfection the combination of two, in principle, separable factors, Leibniz unquestionably drew his inspiration once again from mathematics – as he so often does. Determining the maximum or minimum of that surface-defining equation which represents a function of two real variables specifically requires those problem-solving devices for which the mechanisms of the differential calculus were specifically devised. Unlike the relative mathematical naiveté of the old-line, monolithic, single-factor criterion, the Leibnizian standard of a plurality of factors in nonlinear combination demands the sort of mathematical sophistication that was second nature to him.

3. EPISTEMOLOGICAL IMPLICATIONS

At this juncture we reach an important turning – the transformation of the preceding metaphysical considerations into epistemological ones.

In dealing with the metaphysical standard of perfection and issues of the ethics of creation we are, of course, looking at things from a God's-eye point of view. At that level, one is involved with the God-oriented issue of which among innumerable possible worlds is to be realized. Accordingly, one's concern is with the *ontological* issue of what is to be real among diverse alternatives. The standard of variety-cum-orderliness is to be considered in the light of this metaphysical issue.

But in turning from metaphysics to epistemology we leave this God's-eye perspective behind. Our concern now is not with the God-oriented issue of which existential possibility is to be real (i.e., realized) but with the anthropocentric problem of which among various epistemological possibilities in the phenomenal sphere is to be recognized as actual – i.e., veridical.

Leibniz treats this epistemological issue in one of his most powerfully seminal works, the little tract *De modo distinguendi phaenomena realia ab imaginariis.*[17] How does the golden mountain I imagine differ from the real earthen, rocky, and wooded mountain I see yonder? Primarily in two respects: internal detail and general conformity with the course of nature. Regarding the internal detail of vividness and complexity Leibniz says:

> We conclude it from the phenomenon itself if it is vivid, complex, and internally coherent (*congruum*). It will be vivid if its qualities, such as light, color, and warmth, appear intense enough. It will be complex if these qualities are varied and support us in undertaking many experiments and new observations; for example, if we experience in a phenomenon not merely colors but also sounds, odors, and qualities of taste and touch, and this both in the phenomenon as a whole and in its various parts which we can further treat according to causes. Such a long chain of observations is usually begun by design and selectively and usually occurs neither in dreams nor in those imaginings which memory or fantasy present, in which the image is mostly vague and disappears while we are examining it.[18]

Leibniz proceeds to explicate the nature of coherence as follows:

A phenomenon will be coherent when it consists of many phenomena, for which a reason can be given either within themselves or by some sufficiently simply hypothesis common to them; next, it is coherent if it conforms to the customary nature of other phenomena which have repeatedly occurred to us, so that its parts have the same position, order, and outcome in relation to the phenomenon which similar phenomena have had. Otherwise phenomena will be suspect, for if we were to see men moving through the air astride the hippogryphs of Ariostus, it would, I believe, make us uncertain whether we were dreaming or awake.[19]

He goes on to elaborate the operation of this coherence criterion in considerable detail:

But this criterion can be referred back to another general class of tests drawn from preceding phenomena. The present phenomenon must be coherent with these if, namely it preserves the same consistency or if a reason can be supplied for it from preceding phenomena or if all together are coherent with the same hypothesis, as if with a common cause. But certainly a most valid criterion is a consensus with the whole sequence of life, especially if many others affirm the same thing to be coherent with their phenomena also, for it is not only probable but certain, as I will show directly, that other substances exist which are similar to us. Yet the most powerful criterion of the reality of phenomena, sufficient even by itself, is success in predicting future phenomena from past and present ones, whether that prediction is based upon a reason, upon a hypothesis that was previously successful, or upon the customary consistency of things as observed previously.[20]

Thus, Leibniz lays down two fundamental criteria for the distinguishing of real from imaginary phenomena; the vividness and complexity of inner detail on the one hand and the coherence and lawfulness of mutual relationship upon the other.

Now the interesting and striking fact about this sector of Leibnizian epistemology is *its parallelism* with his ethical metaphysics of creation. In both cases alike, the operative criterion of the real resides in a combination of variety and orderliness. This is certainly no accident. One cannot but sense the deep connection at work here. Let us attempt to illuminate it.

Leibniz's line of thought begins with a theologico-metaphysical application of ethical theory: the doctrine that God will chose for actualization that one among all possible worlds which qualifies as "the best". The implementation of this doctrine, of course, calls for a *metaphysical standard of relative perfection*, a requirement filled by the Leibnizian criterion of lawfulness and variety.

Given this starting-point it is natural to invoke the logical principle of *adaequatio intellectu ad rem* to the effect that, as Spinoza puts it, "the order and connection of ideas is the same as the order and connection of things". Appeal to this principle serves to transmute our metaphsical standard of perfection as used ontologically for bridging the metaphysical division between *possibility* and *reality* into an epistemological standard for bridging the division between *appearance and reality*. In this way, Leibniz shifts the application of the fundamental criterion of variety and orderliness from God's realization-selection of a real among possible worlds to man's recognition-selection of a real among apparent phenomena.

Leibniz's line of thought thus, in effect, exhibits the striking feature of using a logical doctrine, the correspondentist adequation theory of truth and reality, to validate an epistemological coherence theory of truth and reality in terms of the ethico-metaphysical standard of perfection that he views as operative in God's creation choice. This complex and fruitful conjoining of different elements is altogether typical of Leibniz's ingenuity as a philosophic system-builder.

4. LEIBNIZ AS A PIONEER OF THE COHERENCE THEORY OF TRUTH

It is worth noting, as a point of historical interest, that this Leibnizian theory of our knowledge of particular phenomena effectively revives the position of certain of the Academic Sceptics of the

Middle Academy. For Carneades (b. ca. 213 BC) taught essentially this same doctrine that those perceptions which are fully tested – individually plausible, mutually supportive, and systematically coherent – may properly be accepted as veridical. Leibniz adopts just exactly this selfsame view of systematicity as the key criterion of factual truth.

The coherence theory of truth has played a central role in thinking of the Anglo-American idealists from Bradley, Bosanquet, and Joachim to A. C. Ewing and Brand Blanshard in our own day. Moreover, this theory of truth has had a definite appeal for some members of the logical positivist school (O. Neurath, R. Carnap (at one brief stage), and C. G. Hempel (in some passages)). This is not the place to go into details, and I shall simply presuppose familiarity with the theory and its development.[21] But it is relevant to our present concerns to note that the modern coherence theorists articulate a criterion of truth that revolves around exactly the two Leibnizian factors of variety and order.

To see this, consider a passage from the best-known idealist exponent of the coherence theory, the English metaphysician F. H. Bradley:

> There is a misunderstanding against which the reader must be warned most emphatically. The test which I advocate is the idea of a whole of knowledge as wide and as consistent as may be. In speaking of the system I mean always the union of these two aspects, and this is the sense and the only sense in which I am defending coherence. If we separate coherence from what Prof. Stout calls comprehensiveness, then I agree that neither of these aspects of system will work by itself. How they are connected, and whether in the end we have one principle or two, is of course a difficult question .... All that I can do here is to point out that both of the above aspects are for me inseparably included in the idea of system, and that coherence apart from comprehensiveness is not for me the test of truth or reality.[22]

Bradley thus insists emphatically upon conjoining in his own coherence criterion of truth exactly the two Leibnizian factors of order (=coherence) and variety (=comprehensiveness).

These very brief indications should suffice to indicate that Leibniz can be viewed as a pioneer of this line of thought and deserves to be regarded as one of the fathers of the coherence theory of truth.

There is, to be sure, a crucial difference between Leibniz and the English neo-Negelians who espoused the coherence theory of truth at the end of the last century. They are separated by the vast gulf of Kant's Copernican Revolution.

Unlike Leibniz, the modern idealists usually abandoned altogether the traditional correspondence-to-fact idea of truth, and looked upon coherence as affording not an epistemological *criterion* of truth but a logical *definition* of it. They gave up as useless baggage the whole idea of correspondence with an *an sich* reality. This, of course, is a position which Leibniz was unable to take, so that, naturally enough, he remained in the pre-Kantian dogmatic era in which the conception of truth as agreement with an altogether extramental reality was inevitable.

Nevertheless, though on the metaphysical side, Leibniz stays with the dogmatists in his acceptance of an *an sich* reality as the ultimate metaphysical basis of truth, still, on the *epistemological* side, be very definitely foreshadows Kant. Consider the following passage, again from the important little essay *De modo distinguendi phaenomena realia ab imaginariis*:

We must admit it to be true that the criteria for real phenomena thus far offered, even when taken together, are not demonstrative, even though they have the greatest probability; or to speak popularly, that they provide a moral certainty but do not establish a metaphsical certainty, so that to affirm the contrary would involve a contradiction. Thus by no argument can it be demonstrated absolutely that bodies exist, nor is there anything to prevent certain well-ordered dreams from being the objects of our mind, which we judge to be true and which, because of their sccord with each other, are equivalent to truth so far as practice is concerned. Nor is the argument which is popularly offered, that this makes God a deceiver, of great importance . . . For what if our nature happened to be incapable of real phenomena? Then indeed God ought not so much to be blamed as to be thanked, for since these phenomena could not be real, God would, by causing them at least to be in (mutual) agreement, be providing us with something equally as valuable in all the practice of life as would be real phenomena. What if this whole short life, indeed, were only some long dream and we should awake at death, as the Platonists seem to think? . . . Indeed, even if this whole life were said to be only a dream, and the visible world only a phantasm, I should call this dream or this phantasm real enough if we were never deceived by it when we make good use of reason. But just as we know from these marks which phenomena should be seen as real, so we also conclude, on the contrary, that any phenomena which conflict with those that we judge to be real, and likewise those whose fallacy we can understand from their causes, are merely apparent.[23]

The lesson of this passage is clear. In the epistemology of perception we are in no position to implement the correspondentist conception of truth as *adaequatio ad rem* – seeing that we ourselves have no entryway *in rebus*, independent of proceeding on the basis of experience. The distinction between appearance and reality is indeed crucial, but it is for us a distinction strictly to be drawn wholly within the domain of phenomenal reality, and is not a distinction between phenomenal reality on the one hand and noumenal reality on the other.

Thus, while from the God's-eye perspective of his metaphysics Leibniz remains the author of the "System of the Monadology", from the man's eye perspective of the epistemology of particular truths of contingent fact, Leibniz is very much the colleague not only of Kant himself but also of the latter-day idealistic supporters of the coherence criteriology of truth, with the order and system of Leibnizian perfection providing the requisite standard of acceptability.[24]

NOTES

[1] *Discours de mètaphysique,* §2; Loemker, p. 304.

[2] "*Principe de la perfection*", "*Lex melioris*", "*Principe du Meilleur*", "*Principe de la convenance*".

[3] *Discours de mètaphysique,* §2; Loemker, p. 304.

[4] A. O. Lovejoy, *The Great Chain of Being* (Cambridge, Mass.; 1936), p. 179.
[5] GP, Vol. VII, pp. 302–308; Loemker, pp. 486–491.
[6] Loemker, p. 487.
[7] Nelson Goodman, *The Structure of Appearance* (Cambridge, Mass., 1951); 2nd edn. (Indianapolis, 1966); *idem*., "Safety, Strength, Simplicity", *Philosophy of Science*, **28** (1961), 150–151; John G. Kemeny, "The Use of Simplicity in Induction", *The Philosophical Review*, (1953), 291–408.
[8] *Timaeus* 33B; cf. F. M. Cornford's and T. L. Health's comments ad loc.
[9] Recall the stress on generative energy and creative power in the *Enneads* of Plotinus, and passages like "This earth of ours is full of varied life-forms and of immortal beings, to the very heavens it is crowded" (*Enneads* II, 9, §8; MacKenna).
[10] See especially Chap. 11 of Bk. V and Chap. 8 of Bk. IX.
[11] For a more comprehensive treatment of this evaluative dimension of metaphysics, see pp. 230–243 of N. Rescher, *Essays in Philosophical Analysis* (Pittsburgh, 1969).
[12] Edited by G. Dreyfus (Paris, 1958), p. 187.
[13] *Discours de métaphysique,* §6 (Loemker, p. 306); Cf. ibid, §5, and also PNG, §10; *Theodicée,* §208.
[14] *Theodicée,* §208.
[15] Loemker, p. 211.
[16] Loemker, p. 478.
[17] GP, Vol. VII., pp. 319–322; Loemker, pp. 363–365.
[18] Ibid., p. 363–364.
[19] Ibid., p. 364.
[20] Ibid.
[21] For a fuller discussion, including references, see N. Rescher, *The Coherence Theory of Truth* (Oxford, 1973).
[22] F. H. Bradley, "On Truth and Coherence", in *Essays on Truth and Reality* (Oxford, 1914), pp. 202–218 (see pp. 202–203).
[23] Loemker, pp. 364–365.
[24] This essay is a revised version of "Leibniz and the Evaluation of Possible Worlds" first published in the author's *Studies in Modality* (Oxford; Blackwell, 1974).

ESSAY II

THE EPISTEMOLOGY OF INDUCTIVE REASONING IN LEIBNIZ

1. INTRODUCTION

The closing sections of the preceding chapter have examined Leibniz's approach to the epistemology of particular truth in sensory observations. The present chapter will turn to his epistemology of general truth in scientific theorizing. It will emerge that, on both sides alike – the general as well as the particular – considerations of coherentist fit and systematic unity will play the decisive role.

It has been customary to think of Leibniz as first and foremost a metaphysician: the author of the *Monadology* and the founder of the "new system of preestablished harmony". From this perspective, Leibniz's views regarding the theory of knowledge fade into the background. Nevertheless, Leibniz had some extremely interesting ideas in the theory of knowledge, and it is in this somewhat unaccustomed role as an epistemologist that he will be considered in the present discussion.

To be sure, epistemology is negligible if we assume the standpoint that is standard in the articulation of Leibniz's metaphysics. At this level, matters are worked out from a God's eye point of view: everything here turns on the monads and their complete individual concepts as apprehended by God from all eternity *sub ratione possibilitatis*. And then, once the world has come into existence, truth becomes a matter of agreement "with the actual facts" (*adaequatio ad rem*).

But, of course, from the angle of *human* epistemology – from that of the question of how *we* are to find out about how things go in the world – this whole approach affords us no help whatever. We humans do not know the complete individual notions of any existing substances – not even that of our own self or spirit. And this leads straightaway to the issue of the empirical epistemology of the contingent realm – of *how* we men are to extract truths from our experience. This issue, which does not even arise in the *metaphysics* of Leibniz, must nevertheless play a singificant part of his philosophy.

Unlike God, we humans cannot penetrate the realm of contingent fact by means of calculation and reasoning from general principles. Our *only* route into the issue of how things actually stand in this world is through experience. (Reasoning alone cannot carry us beyond the sphere of necessary truths.) And so, from the angle of the practicalities of the human situation Leibniz emerges as a thinker of substantially empiricist learnings.

2. THE EXTRACTION OF GENERAL TRUTHS FROM EXPERIENCE

Let us, then, examine Leibniz's theory as to how *general* truths can be extracted from experience – that is, Leibniz's theory of induction. Experience, as Leibniz repeatedly insists, is inevitably of the particular. The mechanism of perception always involves particular transactions that yield particularized information. With general truths, however, the mediation of reason is crucial: any coloration of generality is always supplied by reason. But how does reason proceed here? How can we make the "inductive leap" from particular experiences to universalized propositions?

Leibniz discusses this question of the methodology of inductive inference briefly in his relatively early (1670) Preface to an edition of *On the True Principles of Philosophy* of Marius Nizolius, and he returns to the issue at greater length in the draft *Elementa physicae* ("On the Elements of Natural Science") written ca. 1682–84.[1]

In this latter work, Leibniz characterizes the proper method of procedure for the validation of general empirical claims as "the conjectural method *a priori*" which "proceeds by hypotheses, assuming certain causes, perhaps without proof (i.e., without any *direct* substantiation), and showing that the things which actually happen would follow from these assumptions".[2]

The deciphering of a cryptogram is Leibniz's favorite illustration of the workings of this method of hypothesis-utilization.

A hypothesis of this kind is like the key to a cryptograph, and the simpler it is, and the greater the number of events that can be explained by it, the more probable it is. But just as it is possible to write a letter intentionally so that it can be understood by means of several different keys, of which only one is the true one, so the same effect can have several causes. Hence no firm demonstration can be made from the success of hypotheses.[3]

The method is conjectural because of its crucial reliance on hypotheses, and *a priori* because of its reliance on fundamental principles whose establishment lies beyond the reach of induction.[4] The aprioricity at issue here indicates not the dispensibility of the empirical, but its incompleteness of insufficiency to the task of establishing, of actually demonstrating the conclusion. As Leibniz writes:

For perfectly universal propositions can never be established on this basis (viz., induction based on the experience of particular cases) because you are never certain in induction that all individuals have been considered. You must always stop at the proposition that all the cases which I have experienced are so. But since, then, no true universality is possible, it will always remain possible that countless other cases which you have no examined are different.[5]

One point clearly emerges from this remarkably Humean passage: there is simply no question of us men ever being in a position to *demonstrate* any general truths of contingent fact.

Induction itself is not enough to afford any sort of "demonstration". The method relies crucially on the conjectural stipulation of theses whose full content lies beyond the reach of realizable experience. And this means that no absolute certainty can be attained with respect to universal theses in the domain of contingent fact. The methodology of a presumption-warranting recourse to methodological principles of order here confines us to the sphere of what is, strictly speaking, merely plausible or probable. But the method can indeed attain the *morally* certain – and so, of course, the *practically* certain which suffices for the guidance of affairs in everyday life.

Some hypotheses can satisfy so many phenomena, and so easily, that they can be taken for certain. Among other hypotheses, those are to be chosen which are the simpler; these are to be presented, in the interim, in place of the true causes. The conjectural method *a priori* proceeds by hypotheses, assuming certain causes, perhaps without proof, and showing that the things which actually happen would follow from these assumptions . . . . Yet I shall not deny that the number of phenomena which are happily explained by a given hypotheses may be so great that it must be taken as morally certain. Indeed, hypotheses of these kind are sufficient for everyday use.Yet it is also useful to apply less perfect hypotheses as substitutes for truth until a better one occurs, that is, one which explains the same phenomena more happily or more phenomena with equal felicity. There is no danger in this if we carefully distinguish the certain from the probable.[6]

The determinative considerations that underlie the attainment of moral certainty in this conjectural sphere are simply the standard epistemological parameters of inductive reasoning: comprehensiveness, generality, uniformity, coherence, simplicity, economy, etc. This last factor, economy, is the pivotal principle of induction for

[*T*]*he simpler a hypotheses is, the better it is.* And in accounting for the causes of phenomena, that hypothesis is the most successful which makes the fewest gratuitous assumptions. Whoever supposes differently by this very fact accuses nature, or rather God, its author, of an unfitting superfluity.[7]

In causal explanation, in particular, we are to seek our explanatory hypotheses by a search for associations reminiscent of Mill's methods of agreement and difference (to put it somewhat anachronistically) applied to the "circumstances" of the phenomena:

But the true method of reasoning from experiments is this – we must resolve every phenomenon into all its circumstances by considering separately color, odor, taste, heat, and cold, and other tactile qualities, and finally, the common attributes of magnitude, figure, and motion. Now if we have discovered the cause of each of these attributes in itself we will certainly have the cause of the whole phenomenon. But if by chance we do not come upon the reciprocal and permanent cause of certain attributes, but only several possible causes, we can exclude those which are not pertinent here. For example, assume two attributes, *A* and *L*, of the same phenomenon and assume that there are two possible causes of *A*, namely, *b* and *c*, and two of *L*, namely, *m* and *n*. Now, if we establish that cause *b* cannot exist along with either *m* or *n*, it follows necessarily that the cause of *A* is *c*. If we can further estabish that *m* cannot exist along with *c*, then the cause of *L* must be *n*. But if it is not in our power to achieve a complete enumeration of possible causes, this method of exclusion will at most be probable. If the effects rather than the causes of a phenomenon are sought, the method is the same; the effects of of the separate attributes will have to the examined.[8]

Underlying such procedures, however, is our commitment to the metaphysical principle of economy and efficiency. Accordingly. induction does not render experience a wholly self-sufficient basis for knowledge, for induction itself rests on certain rational principles. Thus Leibniz writes:

But, you may ask, do we not say universally that fire – that is, a certain luminous, fluid, subtle body, usually flares up and burns when wood is kindled, even if no one has examined all such fires, because we have found it to be so in those cases we have examined? That is, we infer from them, and believe with moral certainty, that all fires of this kind burn and will burn you if you put your hand to them. But this moral certainty is not based on induction alone and cannot be wrested from it by main force but only by the addition or support of the following universal propositions, which do not depend on induction but on a universal idea or definition of terms: (1) *if the cause is the same or similar in all cases, the effect will be the same or similar in all*; (2) *the existence of a thing which is not sensed is not to be assumed*; and finally (3) *whatever is not assumed, is to be disregarded in practice until it is proved*. From these principles arises the practical or moral certainty of the proposition that all such fire burns . . . Hence it is clear that induction in itself produces nothing, not even any moral certainty, without the help of propositions depending not on induction but on universal reason. For if these helping propositions, too, were derived from induction, they would need new helping propositions, and so on to infinity, and moral certainty would never be attained. Perfect certainty can clearly never be hoped for from induction, even with the addition of any aids whatever.[9]

Thus, while the method of hypotheses cannot attain demonstration or absolute certainty, it does enable us to attain to *moral* certainty when we place the experiental data of the induction into the rational framework of certain principles of reason. As Leibniz sees it, induction rests on certain "helping principles" of metaphysical provenience and methodological utility, principles that reflect rules of order provided by reason. Nevertheless, the fact that we are applying these rational principles to the materials of incomplete experience means that this latter becomes determinative and that we can never quite step outside the realm of the probable or plausible in this contingent realm.

The degree of probability that is attained through the conjectural method will hinge on the explanatory power of the conjectural hypotheses at which the method arrives:

Yet it must be admitted that a hypothesis becomes the more probable as it is simpler to understand and wider in force and power, that is, the greater the number of phenomena that can be explained by it, and the fewer the further assumptions. It may even turn out that a certain hypothesis can be accepted as physically certain (*pro physice certa*) if, namely, it completely satisfies all the phenomena which occur, as does the key to a cryptograph. Those hypotheses deserve the highest praise (next to truth), however, by whose aid predictions can be made, even about phenomena or observations which have not been tested before; for a hypothesis of this kind can be applied, in practice, in place of truth.[10]

The actual truth in contingent matters can (presumably) be known by God alone. However, we imperfect inquirers can, do, and should strive to discern those hypotheses which "deserve the highest praise" next to the truth and which we therefore may, in practice, allow to stand in its place. It is predictive efficacy that earns this preferred status because this controls – far more effectively than merely "saving the phenomena" of past observations could ever do – the fit of the hypothesis to all the phenomena: its capacity to accommodate the whole range of observable occurrences.

Leibniz's account of the "conjectural method *a priori*" makes it clear that what is at issue here is exactly what has become known as the hypothetico-deductive method of scientific method. And his account of this method and its *modus operandi* is such that his traditional ranking as a "Rationalist" is highly problematical. It was quite in character, and nowise anomalous, for this "rationalist" to write (already in the early Preface to Nizolius):

> even before the incomparable Lord Bacon of Verulam and other enlightened men recalled philosophy from its airy digressions . . . there were certain barbering alchemists who had sounder and clearer insight into the nature of things than did any philosophaster sitting behind closed doors, bent exclusively over his *hacceitates* . . .[11]

In his emphasis on the dependence of all of our factual knowledge upon observation, his concerns for experimental design, and his views on the nature of hypotheses and the principles for their assessment, Leibniz is, in fact, a committed empiricist.

3. CONCLUDING OBSERVATIONS

In closing, it is deserving of note that the operative factors in the validation of *particular* and of *general* truths are wholly of a piece in Leibniz's epistemology. *Exactly the same sorts of considerations* apply on both the perceptual and the inductive side of our contingent knowledge. For in both cases – as we have now seen – the key lies in the idea of *systematization*, of coherent ordering in the light of rational principles. Systematic fit is everywhere the determinative factor: uniformity, regularity, coherence, simplicity, economy. For Leibniz, systematicity thus emerges as arbiter of truth *throughout* the epistemology of our knowledge of contingent fact. Induction is simply part and parcel of this systematizing venture. This facet of Leibniz's thought gives a curiously modern ring to the *epistemological* discussions of a thinker whose metaphysical writings often sound strange to the modern ear.

To be sure, there is one crucial difference between Leibniz and the moderns. Leibniz endows his recourse to the systematizing impetus with a metaphysical foundation: we are to systematize our cognition because it is to correspond to a reality created by God, the great systematizer.[12] The Kantian justification of systematization as inherent in the *modus operandi* of human reason itself – let alone the pragmatists' defense of systematization as purposively effective for the aims of inquiry – lay far in the future. In this regard, Leibniz is a rationalist metaphysician of the old school.

What is thus remarkable about Leibniz's epistemology of inductive enquiry is that he is neither a doctrinaire rationalist nor an all-out empiricist here, but (as usual) someone who holds to the *via media*. He puts before us a complex and fascinating picture of intricate interplay – even at the level of actual scientific practice – between experience and *a priori* considerations.[13]

NOTES

[1] See Loemker, pp. 125ff. and 277ff, respectively.
[2] Loemker, p. 283.
[3] Ibid.
[4] The method contrasts with "the hypothetical method *a posteriori*, which proceeds from experiments, (and) rests for the most part upon analogies" (Loemker, p. 284).
[5] GP, Vol. IV, pp. 161–2; Loemker, pp. 129–130.
[6] Loemker, p. 283.

[7] Loemker, p. 128.
[8] Loemker, pp. 284–285.
[9] GP, Vol. IV, p. 161; Loemker, pp. 129–130.
[10] GP, Vol. I, pp. 195–196; Loemker, p. 188 (and cf. p. 128). Leibniz's views on probability in the sense of the calculus of probability does not concern us here. For a discussion with references to the literature see Ian Hacking, "The Leibniz-Carnap Program for Inductive Logic", *Journal of Philosophy* **68** (1971), 597–610.
[11] Loemker, p. 124.
[12] This issue will be considered in the next chapter.
[13] This essay is a revised version of the author's paper, "The Epistemology of Inductive Reasoning in Leibniz" in *Theoria Cum Praxi: Akten des III. Internationalen Leibniz Kongress*, Vol. III (Wiesbaden, 1980; *Studia Leibnitiana Supplementa*, Vol. XXI).

ESSAY III

LEIBNIZ AND THE CONCEPT OF A SYSTEM

1. THE CONCEPT OF A SYSTEM

The aim of this essay is to examine the role of the systems concept in Leibniz's thinking. It addresses the questions: whence did Leibniz obtain the idea of systems? How did he develop it? What sort of role did it play in his philosophy?

While the underlying *idea* of what we nowadays call a "system" of knowledge was certainly alive in classical antiquity – with the Euclidean systematization of geometry providing a paradigm for this conception – actual use of the term "system" in this connection is of relatively recent date. In fact, as we shall shortly see, Leibniz was the first thinker to describe himself as having a *system* of philosophy and to attribute the possession of one to others.

To be sure, the underlying idea of a cognitive system unquestionably goes back to classical antiquity. In fact, it has been insisted throughout the history of Western philosophy that men do not genuinely *know* something unless this knowledge is actually systematic. Plato's thesis in the *Theaetetus* that a known fact must have a *logos* (rationale), Aristotle's insistence in the *Posterior Analytics* that strict (scientific) knowledge is a fact about the world calls for its accounting in causal terms, the Scholastic analysis of *scientia*, Spinoza's celebration of what he designates as the second and third kinds of knowledge (in Book II of the *Ethics* and elsewhere), all instantiate the common, fundamental idea that what is genuinely known is known in terms of its systematic footing within the larger setting of a rationale-providing framework of explanatory order. The root idea of system is that of *structure* or *organization*, of integration into an orderly whole that functions as an "organic" unity. And a specifically *cognitive* system is to encompass these desiderata with respect to our knowledge.

A cognitive system is to provide a framework for linking the *disjecta membra* of the bits and pieces of our knowledge into a cohesive unity. A cognitive system is to be a *structured* body of information, one that is organized in accordance with taxonomic and explanatory principles that link this information into a rationally coordinated whole. The functional categories governing this organizational venture are those of understanding, explanation, and cognitive rationalization.

From antiquity to Hegel and beyond, cognitive theoreticians have embraced this ancient ideal that our knowledge should be developed architectonically and should be organized within an articulated structure that exhibits the linkages binding its component parts into an integrated whole and leaves nothing wholly isolated and disconnected.

But while the concept of cognitive systematization is very old, the term "system" itself was not used in this sense until much later. In ancient Greek, *systēma* (from *syn-histēmi*, "to [make to] stand together") originally meant something joined together – a connected or composite whole. The term figures in Greek antiquity to describe a wide variety of composite objects – flocks of animals, medications, military formations, organized governments, poems, musical configurations, among others.[1] Its technicalization began with the Stoics, who applied it specifically to the physical universe (*systema mundi*) – the composite cosmos encompassing "heaven and earth".[2] But the term continued in use throughout classical texts in its very general ordinary sense (which it shared with terms like *syntagma* and *syntaxis*).

The Renaissance gave the term a renewed currency. At first it functioned here too in its ancient applications in its broad sense of a generic composite. But in due course, it came to be adopted by Protestant theologians of the sixteenth century to stand specifically for the comprehensive exposition of the articles of faith, along the lines of a medieval *summa*: a doctrinal *compendium.*[3]

By the early years of the seventeenth century, the philosophers had borrowed the term "system" from the theologians, using it to stand for a synoptically comprehensive and connected treatment of a philosophical discipline: logic, rhetoric, metaphysics, ethics, etc.[4] (It was frequently employed in this descriptive sense in the title of expository books.[5]) And thereafter the use of the term was generalized in the early seventeenth century to apply to such a synoptic treatment of any discipline whatever.[6]

This post-Renaissance redeployment of the term *system* had a far-reaching significance. In the original (classical) sense, a system was a physical thing: a compositely structured complex. In the more recent sense, a system was an organically structured body *of knowledge* – not a mere accumulation or aggregation or compilation of miscellaneous information (like a dictionary or encyclopedia), but a functionally-organized and connectedly articulated expostion of a unified discipline. It was just this sense of "system" that was eventually encapsulated in Christian Wolff's formula of a system as "a collection of truths duly arranged in accordance with the principles governing their connections" (*systema est veritatum inter se et cum principiis suis connexarum congeries*).[7] Moreover, a system is not just a constellation of interrelated elements, but one of elements assembled together in an "organic" unity by linking principles within a functionally-ordered complex of rational interrelationships. The dual application of systems-terminology to physical and intellectual complexes thus reflects a longstanding and fundamental feature of the conception at issue.

A further development in the use of the term occurred in the second half of the seventeenth century. Now "system" came to be construed as a *particular approach* to a certain subject – a particular theory or doctrine about it as articulated in an organized complex of concordant hypotheses, a *nexus veritatum*. (This new usage is particularly marked in Malebranche's treatise *De la Recherche de la verité* [*De Inquirenda veritate libri sex* (Geneva, 1985)], where we find a section "*de novorum systematum inventoribus*".) This is the sense borne by the term in such phrases as "the system of occasional causes" or "the Stoic system of morality".

The prime promoter of this new usage was Leibniz. He often spoke of his own philosophy as "my (new) system" of pre-established harmony, contrasting it with various rival systems. Thus Leibniz contrasts his own *système de l'harmonie préétablie* with the *système des causes efficientes et celui des causes finales* as well as with the *système des causes occassionelles qui a été fort mis en vogue par les belles réflexions de l'Auteur de la Recherche de la Vérité* (Malebranche). He characterizes his own contribution as the *systéme nouveau de la nature et de la communication des substances aussi bien que de l'union qu'il y a entre l'ame et le corps.*[8] Thus as the seventeenth century moved towards its end, the system was now understood as a doctrine or teaching in its fully comprehensive (i.e., systematic) development. And Leibniz was the first who explicitly applied this terminology to a body of philosophical teachings.

2. LEIBNIZ AS SYSTEM BUILDER

The idea of a cognitive system can operate at two importantly distinguishable levels, namely at the level of *concepts* (ideas, conceptions) and at the level of *propositions* (theses, theories, doctrines). From the days of his early *Dissertatio de arte combinatoria* throughout the many decades during which he articulated and refined his projects of a universal characteristic (*characteristica universalis*) and a calculus of reasoning (*calculus ratiocinator*), Leibniz clearly had in mind the prospect of systematizing our knowledge at both of these levels. And this is only to be expected in a thinker of his synoptic and synthetic bent, given the prominence of the Euclidean model of system building, where concept-definitions and basic axioms join together to play a foundational role.

Leibniz sought to implement the idea of a concept-system in the philosophical domain as well, and indeed assigned the system-concept a particularly prominent role in this sphere. For example, he sought, on numerous occasions throughout his career, to provide systematic families of definitions for the fundamental concepts of various domains: ethics, logic, mathematics, natural philosophy, etc. And he certainly thought that the fundamental conceptions of philosophy can be dissected and their interrelationships made evident by analytical processes.

But Leibniz gave no less certain a place to the project of providing for the architectonic construction of philosophy as a system of interrelated principles. In every department of philosophy – natural philosophy prominently included – he saw a complex hierarchy of principles at work:

> But I hold, nevertheless, that we must also consider how these mechanical principles and general laws of nature themselves arise from higher principles and cannot be explained by quantitative and geometrical considerations alone; that there is rather something metaphysical in them . . .[9]

Leibniz's philosophy, more than that of any other thinker, bristles with principles. Their name is legion: we are presented with principles of sufficient reason, of contradition, of perfection, of the identity of indiscernibles, of harmony, and many others adhering to that presentation as a fabric of interrelated principles that is the quintessential hallmark of a system.

As Leibniz sees it, philosophy is – or ought to be – an architectonic structure built up systematically on the basis of various key theses or principles. For Leibniz, a system integrates multiplicity into unity: it combines a rich diversity of content under the accompanying aegis of linking principles. The situation is rather akin to a system of legal rules and regulations, with a hierarchy of principles by reference to which we can achieve the resolution of particular cases.[10] The result should ideally be pretty much a deductive, Euclidean structure whose axiomatic theses and definitions provide the principles from which the whole of the rest can be deduced as consequences. Nevertheless, as we shall see, Leibniz offers his own quite novel variations on these historical themes.

3. WHY SYSTEM?

Just why should Leibniz insist on systematization in philosophy: why should it be that philosophizing should be cast into a systematic form?

This question goes to the very center of Leibniz's "rationalism". A very straightforward line of reasoning is at issue:

(1) The real is a rational order that can, accordingly, only be properly understood on the basis of rational principles.

(2) Rational principles are inherently systematic; by their very nature as rational they have the character of coherent system.

∴ The real can only be properly understood in terms of an appropriate system of rational principles.

It follows from this line of thought that an adequate understanding can only be achieved by means of a system. And this system is the most unified, cohesive, and coherent of possible systems because its basis is the all-pervasive creative agency of God:

> We therefore have the ultimate reason for the reality of essences as well as existences in one being . . . [and] not only the existing things which compose

the world but also all possibilities have their reality through it. But because of the interconnection of all these things, this ultimate reason can be found only in a single source.[11]

4. COGNITIVE vs. ONTOLOGICAL SYSTEMATICITY

The conception of a system has historically been applied both to *things* in the world and to *bodies of knowledge*. It is thus important to distinguish between the *ontological* systematicity (simplicity, coherence, regularity, uniformity, etc.) of the *objects* of our knowledge – that is, between systematicity as a feature pertaining to existing things – and the *cognitive* systematicity of our (putative) knowledge or *information* regarding such things.

In the days of the medieval Schoolmen and of those later rationalistic philosophers whom Kant was wont to characterize as dogmatists, simplicity was viewed as an *ontological feature of the world*. Just as it was held that "Nature abhors a vacuum" – and, more plausibly, "In nature there is an explanation for everything" – so it was contended that "Nature abhors complexity". Kant's "Copernican Revolution" shifted the responsibility for such desiderata from *physical nature* to *the human intellect*. Simplicity-tropism accordingly became not a feature of "the real world", but rather one of "the mechanisms of human thought". Kant acutely observed that what was at issue was a facet not of the teleology of *nature*, but of the teleology of *reason*, responsibility for which lay not with the theory but with the theorizers.

The subsequent Darwinian Revolution may be viewed as taking the process a step further. It transmuted the teleological element. Neither nature nor man's rational faculties are now seen as an ontological locus of simplicity-preference. Rather, its rationale is now placed on a *strictly methodological* basis. Responsibility for simplicity-tropism lies not with the "hardware" of human reason, but with its "software" – i.e., with the procedural and methodological principles which we ourselves employ because we find simpler theories easier to work with and more effective.

On such a latter-day view, it is not that *nature* avoids needless complexity, but that *we* do so – insofar as we find it possible. The parameters of cognitive systematicity – simplicity, regularity, coherence, and the rest – generally represent principles of economy of operation. They are labor-saving devices for the avoidance of complications in the conduct of our cognitive business. They are governed by an analogue of Occam's razor – a principle of parsimony to the effect that needless complexity is to be avoided. Accordingly, cognitive systematicity remains an epistemic factor that is without ontological implications.

Now the important fact from the angle of our present deliberations is that Leibniz stands squarely on the older, dogmatic side of the Kantian divide. For him, systematicity is a God-assured feature of the real, built into the very structure of nature through the criteria of perfection which God deploys in making his creation-selection among alternative possibilities. Thus, in Section 10 of the *Principles of Nature and of Grace*, Leibniz writes:

> It follows from the supreme perfection of God that he has chosen the best possible plan in producing the universe, a plan which combines the greatest variety together with the greatest order; with situation, place, and time arranged in the best way possible; with the greatest effect produced by the simplest means; with the most power, the most knowledge, the greatest happiness and goodness in created things which the universe could allow. For as all possible things have a claim to existence in God's understanding in proportion to their perfections, the result of all these claims must be the most perfect actual world which is possible. Without this it would be impossible to give a reason why things have gone as they have rather than otherwise.[12]

For Leibniz, essence precedes existence and knowledge precedes being: it is because of its possession *sub ratione possibilities* of certain cognitively systemic features that we know, *a priori*, that the real world is an ontological system. And this is the reason why we can monitor the adequacy of attempts to codify our knowledge claims in terms of the extent to which they provide a duly systemalized account of the nature of things.

Since the real world, ontologically speaking, *is* a rational system we ran have a true grasp of it by way of a suitably devised system of rational principles. The principle of truth as *adaequatio ad rem* means that we only have a philosophy adequate to the systematic nature of the real that is itself appropriately systematic. Its systematic nature is accordingly a crucial test and touchstone of the adequacy of a philosophy.

As Leibniz sees the matter systematicity is thus bound to play an absolutely crucial role in philosophy. For its systematic character is the acid test of philosophical adequacy once the matter of "saving the phenomena" – of adequacy to the elemental facts of the matter – is provided for. Smoothness of accommodation of the presystematic givens of our existence – their being woven into a seamless fabric of rational principles – is the determinative mark of adequate philosophical system. And it is his sense of having devised a system that does the job to an extent greater than any rival alternative that is the basis of Leibniz's confidence in the ultimate correctness of "the system of pre-established harmony".

5. SYSTEM AND INFINITE COMPLEXITY

As we have seen, the idea of system operative in Leibniz's thought involves the interweaving of two inseparable threads: diversity (variety, richness of content) on the one hand and economy (simplicity, unity, elegence) on the other. Systematicity indicates the presence of unity amidst variety: the perfect system exhibits an infinite richness of content amidst an elegant and economical order of operative principles. And these two elements stand in an inextricable conjunction:

There is always a principle of determination in nature which must be sought by maxima and minima, namely that a maximum effort should be achieved with minimum outlay, so to speak.[13]

It is clear that these two aspects of *system* are exactly the same as those of *perfection* that – according to Leibniz – qualifies possibilities for actualization in the eyes of God. The world is ontologically systematic precisely because rational systematicity is itself the standard of existence. Significantly, it is the calculus that provides Leibniz with his model for how this sort of thing is to work.

Historically, a *deductive* (Euclidean) system has always been taken as the quintessential paradigm of a cognitive system capable of yielding an endless variety of results on a cognitive basis. This geometric model of cognitive structure holds that the organization of knowledge must proceed in the following manner. Certain theses are to be basic or foundational: like the axioms of geometry, they are to be used to justify other theses without themselves needing or receiving any intrasystematic justification. Apart from these fundamental postulates, however, every other thesis of the system is to receive justification of a rather definite sort. For every nonbasic thesis is to receive its explanation along an essentially linear route of *demonstration* (or derivation or inference) from the basic theses that are justification-exempt or self-justifying. There is a step-by-step, recursive process – first of establishing certain theses by immediate derivation from the basic ones, and then of establishing further theses by sequential derivation from already established theses. Systematization proceeds in the manner characteristic of axiomatic systems.

In the setting of this Euclidean model of cognitive systematization, every (nonbasic) established thesis is ultimately connected to certain basic theses by a linear chain of sequential inferences. These axiomatic theses are the foundation on which rests the apex of the vast inverted pyramid that represents the total body of knowledge. On this approach to cognitive systematization, one would, with J. H. Lambert, construe such a system on analogy with a building whose stones are laid, tier by successive tier, upon the ultimate support of a secure foundation.[14] Accordingly, the whole body of knowledge obtains with regard to its justificatory structure the layered make-up reminiscent of geological stratification: a bedrock of basic theses surmounted by layer after layer of derived theses, some closer and some further removed from the bedrock, depending on the length of the (smallest) chain of derivation that links any given thesis to the basic ones.

It is almost impossible to exaggerate the influence and historical prominence this Euclidean model of cognitive systematization has exerted throughout the intellectual history of the West. From Greek antiquity through the eighteenth century it provided an ideal for the organization of information whose influence was operative in every field of learning. From the time of Pappus and Archimedes and Ptolemy in antiquity to that of Newton's *Principia* and well beyond into modern times the axiomatic process was regarded as the appropriate way of organizing scientific information. And this pattern was followed in philosophy, in science, and even in ethics – as the *more geometrico* approach of Spinoza vividly illustrates. For over two millennia, the Euclidean model has provided the virtually standard ideal for the organization of knowledge.

With Leibniz, however, we see a new turning. For him, the ideal model of a cognitive system was provided not by the geometry of the ancient Greeks but by the physics of seventeenth-century Europe. In his view, the calculus was not just a convenient mechanism for solving mathematical problems, it provided a new model of rational systematization, for implementing that "principle of determination in nature which must be sought by maxima and minima". Such an instrument provides, as Leibniz saw it, a tool for discerning, amidst an infinite variety of diverse phenomena, those operative principles (for example, Snell's law in optics) through which the desiderata of rational economy are instituted in the nature of things.

With a Euclidean axiomatization we have a finite basis of elements (axioms and definitions) from which these are extracted by finite deductive processes. With the calculus – and especially the calculus of variations – we are put into a position of being able to survey an infinite range of alternatives and to discern amidst a literally endless variety of possibilities those particular determinatives indicated by the principles of rational economy. Here, with this system-oriented resort to the mechanisms of the calculus, we once again see at work in the thought of Leibniz a Renaissance-inspired bursting of the classical bonds.

NOTES

[1] Much of the presently surveyed information regarding the history of the term is drawn from the monograph by Otto Ritschl, *System und systematische Methode in der Geschichte des wissenschaftlichen Sprachgebrauchs und der philosophischen Methodologie* (Bonn, 1906). Further data are given in the review of Ritschl's work by August Messer in the *Gottinger gelehrte Anzeigen*, Vol. 169 (1907), No. 8. See also Aloys von der Stein, "Der Systembegriff in seiner geschichtlichen Entwicklung", in A. Diemer (ed.), *System und Klassification in Wissenschaft und Dokumentation* (Meisenheim am Glan, 1968).

[2] See Theodore Ziehen, *Lehrbuch der Logik* (Bonn, 1920), p. 821. The foundation of the Stoic's approach lies in Aristotle's *De mundo*. But contrast also Sextus Empiricus, *Outlines of Pyrrhonism*, III, 269, which speaks of the *systema* of the rules of art, or again *ibid*. II, 173, which speaks of the *systema* (=collectivity) of the propositions of a syllogism.

[3] Thus, Du Cange, *Glossarium mediae et infimae latinatatis* (Paris, 1842): *Systema, proprie compages, collectio. Hinc astronomis pro mundi constitutione et forma usurpatur. Theologis vero pro complexu articulorum fidei.* The term gradually drove its rival *syntagma* from the field in this latter sense.

[4] Thus, Bartholomaeus Keckerman (d. 1609) wrote in his treatise *Systema logicae tribus libris adornatum* (Hanover, 1600), that what is at issue is the

whole organized body of logical precepts. He explained that the term *logic*, like that for every art, stands for two things: The practical skill (*habitus*) on the one hand, and the systematic discipline on the other: *primo pro habitu ipso in mentem per praeceptu et exercitationem introducto: deinde pro praeceptorium logiocorum comprehensione seu systemate* .... (Quoted in O. Ritschl, op. cit., p. 27.) Keckerman's later (1606) handbook of logic appeared under the title *systema minus*. His contemporary Clemens Timpler (d. *ca.* 1625) wrote in his *Metaphysicae systema methodicum* (Hanover, 1606) that in an exposition that is ordered and structured according to proper methodological principles: *systema non confusum et pertubatum, sed bene secundum leges methodi ordinatum et dispositum.*

5 More than 130 titles of this sort published during the seventeenth century are listed in Ritschel (op. cit.). Some examples: Johann Heinrich Alsted, *Systema mnemonicum duplex* (Frankfurt, 1610); Nicas de Februe, *Systema chymicum* (Paris, 1666 [in French]; London, 1666 [in English], Richard Elton, *Systema artis militaris* (London, 1669). (For details see O. Ritschl, op. cit.)

6 There is no entry for *system* in the *Lexicon Philosophicum* of Rudolf Goclenius (Frankfurt, 1623). But in that of Johann Micraelius (Stettin, 1653) the term is explained in its literary sense as a systematic exposition.

7 *Logic*, Sect. 889; cited in Theodor Ziehen, op. cit., p. 821.

8 Cf. Loemker, p. 587.

9 Loemker, p. 409.

10 Compare Leibniz's description of his systematization of Roman Law in his letter to Arnauld of 1671 (GP, Vol. I, p. 73).

11 Loemker, p. 489.

12 Loemker, p. 639.

13 Loemker, p. 487.

14 J. H. Lambert, *Fragment einer Systematologie* in J. Bernouilli (ed.). *Johann Heinrich Lambert: Philosophische Schriften* (2 vols; reprinted Hildesheim, 1967).

ESSAY IV

LEIBNIZ ON THE INFINITE ANALYSIS OF CONTINGENT TRUTHS

1. INTRODUCTION

Leibniz taught that not only *necessary* truths are "analytic", but so also *contingent* ones are. Their analysis, however is infinite and can be effected only by God, who alone can carry through the *a priori* demonstration that exhibits the truth of a contingent proposition.

But how is the mere difference between finite and infinite analyticity able to bear the great burden that Leibniz puts upon it. Why should the mere length or complexity of the analyzing demonstration make for that *toto caelo* difference at issue in the dichotomy between the necessary and the contingent? Given that *both* contingent and necessary truths are demonstrable *a priori*, that both are altogether certain, what justifies all this Leibnizian to-do about there being so profound a difference here? How can something that seems (on first view at any rate) to be a trivial difference – the matter of the mere length of an analysis – support the weight of the most important single distinction of Leibnizian metaphysics?

To elucidate this issue, we must consider not merely *that* a distinction between finite analysis and infinite analysis is at issue, but also *how* this difference operates. For at bottom the matter turns not just on the *length* of the analysis, but on the underlying, length-explanatory differences in the *mode* of analysis.

2. ANALYSIS

Let us begin by considering how a Leibnizian "analysis" works in the case of a necessary proposition. Take "all men are rational animals" for example. The idea here is that terms at issue "men" and "rational animal" can be resolved into conceptual constituents as per the following exfoliation. For "man" we have:

> finite being, of human configuration, capable of bisexual self-reproduction, endowed with reason, possessed with a capacity for moral agency.

For "rational animal" we obtain:

> finite being, self-reproducing, endowed with reason.

Now, with a necessary proposition, it eventuates that the obtaining of the relationship represented by the proposition at issue can be settled by straight-forward deduction from the concept-decomposition of the items at issue. Specifically, "All *A*'s are *B*'s will be necessary if every item in the concept-decomposition register of *B* is also present in the concept-decomposition register of *A* exactly as is the case with the preceding example. The issue here is one of explicit property-containment determinable through the use of definitions.

With contingent propositions, however, a very different story is involved.

Within the framework of the present discussion, we shall focus on contingent propositions about *individual substances*. And this is no serious limitation. As is well-known, Leibniz maintains that *all* other contingent propositions are reducible to such (in ways that need not concern us within the context of our present purposes, but which will be considered in detail in the next chapter).

To be sure, an individual does not (exactly) have a *definition* but rather a "complete individual concept" (a phrase for which we shall use the abbreviation CIC). This view is deeply embedded in Leibniz's creation-metaphysics. For prior to the creation, when the substance at issue does not yet exist at all, its concept is to be found present, *sub ratione possibilitatis*, in the great inventory of possibilities that is part of the stock of God's mind.

Now we might as well follow Leibniz's own example and use *people* as paradigm individual substances. For, in general, the mid-sized physical objects that we encounter in human experience are not – on Leibniz's view – substances at all, but aggregates of substances that individually lie beneath the threshold of perception. Only in the case of people, is one individual substance (viz. the unifying and individuating mind or spirit) so closely associated with an experienced aggregate that we can view it as qualifying as a substance. Consider, then, a paradigm contingent proposition about a human individual:

Caesar crossed the Rubicon.

To determine the truth of this proposition, all that needs to be done is for God simply to check that "Rubicon crossing" is present within the vast constellation of predicates that represents the complete individual concept (CIC) of Julius Caesar:

Caesar ~ {. . . , crosses Rubicon, . . .}

Seemingly what is at issue here is just a matter of a slightly more complicated checking process: finding the needle in an infinite rather than finite haystack.

To be sure, this is a process that can be carried out only by God, since he alone knows the CICs of individual substances – as we certainly do not (indeed not even in the case of that substance regarding which we have optimal information, namely ourselves). But still, the general principle is exactly the same: it is just a matter of looking something up on a list. Why should the mere length of the list make all that much difference?

This perspective, however, overlooks something very fundamental. Consider the finite case of an analytical concept-decomposition:

$$A \sim \{C, D, E\}$$

How is it determined that this correspondence in fact obtains – and so how is it to be known that the reductive analysis based on this correspondence is correct? The answer is simple. We know this merely and solely by the *definition* of the concept A and perhaps also the further definitions of the concepts that figure there. Thus, no serious complications arise here. The correspondence between "the (concept) A at issue" and its analytical decomposition is a *purely theoretical* linkage that obtains in view of the definitions of the terms at issue. It holds "in every possible world", and is established simply by the sort of *Wesenschau* at issue in discerning the definitional content of concepts.

In contingent case, however, the matter stands on a very different footing.

Consider the complete individual concept of a particular contingent individual, Julius Caesar, for example:

$$\text{Julius Caesar} \sim \{\phi_1, \phi_2, \ldots, \phi_j, \ldots\}$$

There are two ways of looking at this:

(1) Given that we are dealing with an individual of *this* description – with this particular constellation of properties – and so with the property of ϕ_j in this particular descriptive context – what is it that qualifies this so-described individual for existence, that is, for membership in this, the actual world?

But this issue can also be looked on the other way round:

(2) Given that we are dealing with an *existing* individual – our Julius Caesar as a member of this, the actual world – what is it that constrains *this* existing substance to have ϕ_j *in this particular descriptive context* $\{\ldots \phi_j$

. . .} rather than some contrary (or contradictory) of ϕ_j?

The first way of looking at the issue asks why an individual answering to a *given* description should exist. The second asks why an (*ex hypothesi*) *existing* individual should answer to a description of a certain sort. Approached in the second way, the issue is this: why should it be that the possible individual substance answering to the identifying description (complete individual concept)

$$s \sim \{\phi, \psi, \ldots, \chi, \ldots\}$$

should have just exactly χ at its position in this context (rather than some contrary or contradictory thereof). That is, just how is it that this particular way of filling in the (otherwise unchanged) description is optimal in maximizing perfection. Specifically, if some predicate χ' distinct from χ were present in the preceding substance-description, this would render it a substance s' whose presence in the world – *mutatis mutandis* – would be perfection-reducing. (And this indeed holds for *every* one of the infinitely many properties comprising the CIC of s.)

This second perspective accordingly implements in the case of truths about contingent existents the basic predicate-in-concept principle, regarding which Leibniz writes to Arnauld:

For *there must always be some foundation for the connection of the terms of a proposition,* and this *must be found in their concepts.* This is my great principle, with which I believe all philosophers should agree . . .[1]

From the angle of the ontology of the Leibnizian creation-ethic, the former way of looking at it is appropriate. But from the angle of the criteriology of contingent truth (where the question of existence can already be viewed as settled – for there can be a realm of contingent *truth* only where there is a domain of contingent *fact*) the second way of looking at it is clearly apposite.

Now it is exactly this second issue of showing that the predicate does properly belong within its context in the defining concept of an existing individual that is the task of analysis. For here, as elsewhere, it is the mission of analysis to show that the predicate is continued in (the content of) the subject: *praedicatum inest subjecto*. Of course, to determine that an *existing* substance *s* is at issue we must ascertain that not only the given resolution with respect to the property χ is perfection-optimizing, but must carry this through seriatim, item by item, for every property in *s*'s CIC. This is why, as Leibniz writes:

A true contingent proposition cannot be reduced to identical ones, but is proved by showing that as the analysis is continued further and further, it constantly approaches identical ones, though never reaches them. Therefore it is God alone, who encompasses the entire infinite in his mind, who knows all contingent truths with certainty. So the distinction between necessary and contingent truths is the same as that between lines that meet and asymptotes, or between commensurable and incommensurable numbers.[2]

But how can such an analysis proceed? It is Leibniz's recourse to the Principle of Perfection that enable us to understand what he intends when he speaks of the contingent truths as analytic, but yet requiring an infinite process for this analysis. A given proposition concerning a contingent existence is true, and its predicate is, indeed, contained in its subject, if the state of affairs characterized by this particular inclusion involves a greater amount of perfection for the world than any other possible state; i.e., if the state of affairs asserted by the proposition is one appropriate to the best possible world.

All contingent propositions have sufficient reasons, or equivalently have *a priori* proofs which establish their certainty, and which show that the connection of subject and predicate of these propositions has its foundation in their nature. But it is not the case that contingent propositions have demonstrations of necessity, since *their sufficient reasons are based on the principle of contingence i.e., on is best among the equally possibly alternatives . . .* [3]

A truth of fact is such that the state of affairs it asserts is one belonging to the best of all possible worlds, hence its analysis, which consists in showing that this is indeed so, requires an infinite process of comparison. Accordingly, what is at issue in the analysis of a contingent truth is not simply by the substitution of world-indifferently defined terms (as in the necessary case). What is needed is an infinite comparison process that shows that putting a particular predicate at the CIC place in question will effect the solution of a certain maximizing problem – a maximization which shows a possible substance with a certain particular descriptive constitution to be apposite for the best of possible worlds.

It is thus via the infinite comparison demanded by the Principle of Perfection that an infinite process is imported into the analysis of a truth dealing with contingent existence. The Principle of Perfection represents the maximum principle that furnishes the mechanism of God's resolution of the problem of a creation choice among the infinite, mutually exclusive systems of compossibles.

3. CALCULUS AS THE INSPIRATION OF INFINITE ANALYSIS

The Principle of Perfection accordingly provides the explanation of, and the mechanism for, the infinite analysis of contingent truths. A given proposition of the contingent type is true, and its subject includes its predicate, if the state of affairs characterized by this inclusion allows of greater perfection for the world than any other state of affairs. A truth of fact is such that the state of affairs that it asserts is one belonging to the best of all possible worlds – and thereby the actual one. And its analysis, which consists in showing that this is indeed so, calls an infinite process of comparison whose methodology approximates that of the infinitesimal calculus.

For Leibniz, the infinite analysis of a contingent truth is to be conceived of an analogy with the infinitistic comparison problems of the calculus of variations, whose mission is the resolution of problems of maximization subject to constraints. This branch of mathematics, which numbers Leibniz among its founders, handles problems such as selecting from among the infinite number of equiperimetric triangles that of maximum area or from among the infinite number of curves that of fastest descent.

Exactly this sort of process is at issue in the implementation of the Principle of Perfection. Under the auspices of this principle, the divine mind solves such problems as selecting that possible world with the maximum of perfection, or that possible Adam whose existence entails the greatest number of desirable consequences.[4]

In line with the Principle of Perfection, the actual world as a whole is as perfect as possible and each of its parts is itself, in turn, as perfect as possible. There are not, as with Descartes, partial imperfections compensated for by the perfection of the whole. Each part of the world aids in the maximization of perfection by contributing the maximum of perfection that is, under the circumstances, possible for it.

The Principle of Perfection accordingly resembles physical principles of a familiar kind. Leibniz himself makes much of such as the optical principles of least (or greatest, as the case may be) time and distance, and the principle of least action. The Principle of Perfection also specifies that in nature some quantity is at a maximum or a minimum, an idea which Leibniz often illustrates with the remark that in nature a drop of water will, if undisturbed, take the form of a sphere, enclosing a maximal volume with a surface of given area. In other words, we have a "minimax" or "extremal" principle.[5] Like the others, it requires techniques analogous to those of the calculus of variations. And the principle was, in fact, suggested to Leibniz by the analogous mathematical considerations. The background here is important for a proper understanding of his thought.

The remark is due to Hero, and is also to be found in the Ptolemaic corpus, that in traveling from one point to another via a plane mirror, a ray of light takes the shortest path.[6] In Leibniz's day a generalized version of this principle was beginning to find a place in optics. In about 1628 Fermat had developed a principle of least time which, together with a method of maxima and minima which anticipated the calculus, he used to deduce the laws of reflection and the newly discovered law of refraction.[7] Leibniz ardently espoused this principle, and reproached Descartes with having used, in accordance with the Cartesian program, a more clumsy mechanical method in the derivation of Snell's law of refraction, instead of the more elegant and fundamental principle of least time or distance.[8]

The mathematical problem of maxima and minima led Leibniz to extend Fermat's investigations, and resulted in invention of the differential calculus, but this does not concern us here. However, the optical minimum principle of least time also absorbed much of Leibniz's time and interest, and he generalized it in an important way. This generalization of Fermat's principle of least time (or equivalently in the case of a constant speed, of least distance) is to the effect that there need not be, in the usual transmission phenomena, a minimization of time. As Leibniz rightly points out, there might be a maximization, for example, in the case of a concave mirror.[9] But this does not undermine the general principle, for Leibniz remarks that the mathematical method for finding maxima and minima which he developed – finding a zero of the first derivative – yields both maxima without discriminating between them.[10]

Inspired by its success in optics, Leibniz sought ardently to extend the applicability of minimax principles by means of a general principle that, in all natural processes, some physical quantity is at a maximum or a minimum. He felt he had found a principle which cuts across the particular laws of physics, and clearly demonstrates the general interconnection of things. Here was a powerful unifying rule for the multitude of particular natural laws, giving coherence to natural science and showing clearly the economy in nature. He could draw upon pure mathematics, mechanics, optics, and dynamics for illustration of his principle.[11]

Throughout problems of this sort, the object is to find one among an infinite number of alternative paths that achieves an extremization (minimization or maximization) of some specified characteristic (time, distance, and so one). What is at issue in such physical problems is an infinite comparison process leading to selection of an optimal alternative.

In his doctrine of contingence, perhaps more heavily than in any other part of his philosophy, Leibniz the philosopher is indebted to Leibniz the mathematician. The logic underlying this doctrine stems entirely and directly from Leibniz's mathematical investigations:

> There is something which had perplexed me for a long time – how it is possible for the predicate of a proposition to be contained in (inesse) the subject without making the proposition necessary. But the knowledge of Geometrical matters and especially of infinitesimal analysis, lit the lamp for me so that I came to see that notions too can be resolvable in infinitum.[12]

> At length some new and unexpected light appeared from a direction in which my hopes were smallest – from mathematical considerations regarding the nature of the infinite. In truth there are two labyrinths in the human mind, one concerning the composition of the continuum, the other concerning the nature of freedom. And both of these spring from exactly the same source – the infinite.[13]

4. A METAPHYSICAL CALCULUS OF PERFECTION-OPTIMIZATION

The predicate-in-subject analysis of a contingent truth regarding actual existence accordingly rests on a metaphysical calculus of perfection-optimization that shows that this particular resolution of the predicate-containment question is optimistic – that the subject with this particular predicate, as part of its descriptive constitution, belongs to the best possible world.

Leibniz insists in the important letter to Arnauld of 14 July 1686, "there must always *be some foundation* for that connection of the terms of a (true) proposition which must be present in their concepts", a circumstance that affords the reason why "every individual substance expresses the whole universe according to its own way and under its particular aspect, or, so to speak, according from the point of view from which it is regarded".[14]

To show that a predicate of an existing substance properly belongs within its CIC context – given that this CIC is construed as characterizing an *existing* substance – requires an infinite comparison process. This introduces an infinitistic element into the demonstration that *praedicatum inest subjecto* – i.e., into the analysis in question. The analysis turns infinite at the exact point of assuring that one is dealing with "the right" concepts – a problem that does not arise in the case of necessary proposition, because here it is settled by the operative definitions.

The complete individual concept (CIC) of a substance is itself a structure of infinite complexity: it involves specifying an infinity of properties. Hence, when it is shown that putting ϕ at the indicated CIC place in the overall context $\{\ldots, \psi, \phi, \chi, \ldots\}$ is a maximizing solution (i.e., a perfection-optimizing one), there remains the question of showing that each of the various *other* co-present properties, ψ, χ, etc., must also go at its proper place within this CIC context. Only in this asymptotic limit can one thus manage to validate the entire CIC at issue as appropriate to the *existing* substance in question. That is, only in the limit do we converge on that complete and actually substance-individuating description definitive of the individual of which we were (from the very outset) to show that ϕ belongs among *its* defining features.

It is in this sense that one should construe Leibniz's assertion from the *General Investigations*:

> A true contingent proposition cannot be reduced to identical propositions, but is proved by showing that if the analysis is continued further and further, it constantly approaches identical propositions, but never reaches them. Therefore it is God alone, who grasps the entire infinite in his mind, who knows all contingent truths with certainty. So the distinction between necessary and contingent truths is the same as that between lines that meet and asymptotes or between commensurable and incommensurable numbers.[15]

The issue, that is, is one of getting to a substance-description that approaches a definite value only asymptotically in the infinite limit. Only asymptotically do we converge on that substance-identifying characterization of which we can say at last: *praedicatum inest subjecto*, so that the overall task of analysis is actually achieved.

For Leibniz, then, this over-all task of demonstrating *a priori* the (infinite) analyticity of a contingent proposition is simply another side of the coin of that of showing a substance of a certain description to belong to the real world (i.e., the perfection-optimizing one). The infinite analyticity that is determinative of truth and the perfection-optimization that is determinative of real existence are, for Leibniz, simply different ways of looking at the workings of one selfsame process.

5. CONCLUSION

If this perspective is correct, then there is a profound difference in "analysis" between the finite and the infinite case. To be sure, in both cases the *object* of the enterprise is precisely the same: to show that the predicate is properly enclosed within the concept of the subect. But in the case of necessary truths, this is a finite process that requires recourse to definitions alone, while in the contingent case it is something that cannot be settled by definitions (*Wesenschau*) alone, but requires the infinitistic comparisons of a maximization process that alone can assure the connection between essence and existence that is the basis for the contingent truth – thus issuing in a task of infinite complexity, involving procedures that can be carried out by God alone. To quote Leibniz:

> Therefore human beings will (not) be able to comprehend contingent truths with certainty . . . . There can be circumstances which, however far an analysis is continued, will never reveal themselves sufficiently for certainty, and are seen perfectly only by him whose intellect is infinite. It is true that as with asymptotes and incommensurables, so with contingent things we can see many things with assurance, from the very principle that every truth must be capable of proof . . . . But we can no more give the full reason for contingent things than we can constantly follow asymptotes and run through infinite progressions of numbers.[16]

This perspective, then, solves the question that we had posed at the outset. It shows that the process of analysis – i.e., of embedding the properties in question in the concepts of the subjects at issue *viewed as existents* turns infinitistic at just the point that separates the contingent from the necessary – the point of the issue of existence in this, the actual world, which is to say the best possible one. For at this point it involves a detour through the divine calculus of optimization or perfection-maximization that affords, for Leibniz, the mechanism of God's creative procedure.

In Leibniz's analysis of contingence, the mathematical analogy of the calculus and especially the calculus of variations is centrally important. The basic point, of course, which must never be lost sight of, is that contingent propositions are true solely because their truth-status derives from God's free choice to implement the principle of perfection. But this choice enmeshes the whole process in the operation of an infinite analysis. And this analysis involves infinite convergent processes, analogous to those of the calculus, but so complex that only God *can* carry them out, and alone *need* do so.

NOTES

[1] GP, Vol. II, p. 56 (Loemker, p. 337).

[2] Couturat, *Opuscules*, p. 388.

[3] GP, Vol. IV, pp. 438–439 (italics supplied).

[4] "The true reason why this thing which is better than that exists is to be subsumed under the free decrees of the divine will, of which the primary one is the decision to do everything in the best possible way, as seems wisest. Therefore it is occasionally permitted that a more perfect to excluded in favor of a less perfect; nevertheless in the sum that way of creating the world is chosen which involves more reality or perfection, and God works on the model of the master Geometer who in problems brings forth the best construction. Therefore all beings, insofar as they are involved in the first Being, have, above and beyond bare possibility, a propensity toward existing in proportion to their goodness, and they exist by the will of God unless they are incompatible with still more perfect possibilities" (GP, Vol. VII, pp. 309–310, notes; cp. 303–304).

[5] See the *Tentamen Anagogicum* (GP, Vol. VII, pp. 270–279) on the minimax principles and their relation to the Principle of Perfection. The reader interested in the principle of least action in Leibniz is referred to the sixteenth note appended to Couturat's *Logique*, and to the Appendix to M. Gueroult's *Dynamique et metaphysique leibniziennes* (Paris: J. Vrin, 1934).

[6] See E. Mach, *Science of Mechanics*, trans. P. E. B. Jourdain (LaSalle, Ill.: Open Court Publishing Co., 1915), p. 518, and GP, Vol. VII, p. 274 (Loemker, p. 479).

[7] The law of refraction is due to Snell (1612). Cf. GP, Vol. VII, pp. 273–738 (Loemker, pp. 478–484).

[8] GP, Vol. VII, p. 274 (Loemker, pp. 479–480).

[9] Ibid., pp. 274–275 (Loemker pp. 480–481).

[10] Ibid., p. 275 (Loemker, p. 481).

[11] See, passim, in the *Tentamen Anagogicum*, GP, Vol. VII, pp. 270–279 (Loemker, pp. 477–484).

[12] Couturat, *Opuscules*, p. 18. Compare the important essay "On Freedom" in Loemker, pp. 263–266.

[13] Couturat, *Logique*, p. 210, notes.

[14] GP, Vol. II, pp. 56–57.

[15] Couturat, *Opuscules*, p. 388.

[16] Ibid., pp. 388–389.

ESSAY V

LEIBNIZ ON INTERMONADIC RELATIONS

1. INTRODUCTION

His theory of relations represents a crucial component of Leibniz's philosophy. It is one of those many points of fertile concurrence where logic and metaphysics come together in fruitful symbiosis. But the theory confronts many problems. One of the gravest of these is the question: Are intermonadic relations real, or are they matters of mere seeming? It is sometimes said that for Leibniz, relations are mere creatures of the mind: That they are "our own creation" and "not in nature", etc. This is indeed the doctrine of Spinoza,[1] and is approximated by such nominalist and materialist thinkers as Gassendi[2] and Hobbes.[3] But it is by no means the theory of Leibniz. Nothing could be further from the truth than Bertrand Russell's contention that on Leibniz's theory a relation is merely an "ideal thing" and that:

> If he were pushed as to [the nature of] this "ideal thing", I am afraid he would declare it to be an accident of the mind which contemplates [it].[4]

The aim of this discussion is to examine and clarify Leibniz's position on this pivotal issue of the reality and the nature of intermonadic relations.

2. THE CRUCIAL ROLE OF RELATIONS IN INCOMPOSSIBILITY

His view of intermonadic relations is crucially likened to Leibniz's theory of possible worlds. Since the complete individual concept of a Leibnizian substance embraces the specification of literally every facet of its career;[5] it involves all details of the relation of this substance to others.[6] But now suppose that:

(1) Possible substance #1 has the property P and also has the feature that there is no substance having property Q to which it (#1) stands in the relationship R.

(2) Possible substance #2 has the property Q and also has the feature that every substance having the property P stands in the relationship R to it (#2).

These two substances are patently incompatible (on logical grounds). God might realize #1 or He might realize #2, but He cannot possibly realize both of them. (It is a fundamental tenet of Leibniz's philosophy that even omnipotence cannot accomplish the impossible.) Substances which clash in this way are characterized by Leibniz as *incompossible*. All of the substances comprising any possible world must, of course, be mutually *compossible* in the correlative sense: otherwise they could not possibly coexist within one shared world-environment.

The purely logical fact that different possibilities can be mutually incompatible – that not all possibilities will be *compossible* in admitting of concurrent realization – plays a crucial role in Leibniz's metaphysics. It yields the reason why God must unavoidably choose between *alternative* schemes of things, between different possible worlds. The need for such a choice is crucial to God's status as a moral agent. As Leibniz wrote the day after his meeting with Spinoza in 1676:

> If all possibles existed, no reason for existence would be needed and possibility alone would suffice. Therefore there would be no God save insofar as He is possible. But such a God as the pious believe in would not be possible if the opinion of those is true who hold that all possibles exist.[7]

The prospect of incompossible worlds is thus a logical circumstance that has profound philosophical and theological implications. And the crucial role of relations in Leibniz's system emerges against this background. For it transpires that *if substances were not mutually interrelated there would be no prospect of their incompossibility*. Leibniz's clear insistence that substances have relational properties – as with "Adam is the father of Cain" – is motivated by the consideration that in this way alone can substances be incompossible, as the actual Adam is incompossible with a Noah characterized as (*inter alia*) "the father of Cain".

It is clear now how one must proceed in endeavoring to answer, on Leibniz's behalf, the question: How can the complete individual notions of two individually possible substances a and b be so constituted as to render them mutually incompossible? One begins with two crucial observations:

(1) "being related by R to b [i.e., to *any* substance answering to b's description]" is one of the relational properties of a, i.e., a has $(\lambda x)xRb$.

and further

(2) "not standing in the relation R to a [i.e., to *any* substance answering to a's description]" is one of the relational properties of b, i.e., b has the property $(\lambda x){\sim}aRx$

These two are clearly incompossible, seeing that (1) entails aRb whereas (2) entails ${\sim}aRb$. Two individuals x and y can only be pairwise incompossible if their complete individual concepts are such that, in some case where one of them has the relational property of R-ing the other, this other lacks the relational property of being-R'd by the former. Incompossibility must ride upon interrelationships.

Consider two possible substances defined via their complete individual notions:

a: $A_1, A_2, A_3, \ldots$
b: $B_1, B_2, B_3, \ldots$

How can an incompatibility arise? Clearly only if we can extract from the facts regarding *a*'s possession of the attributes A_1, A_2, etc. that "There is no substance *x* that answers to *b*'s description". But this conclusion can follow from the description of *a* only if this description somehow incorporates relational information about substances distinct from *a*.

Thus, while Leibniz emphatically denies that relations are somehow substantial entities in their own right, he is not in a position to deny that they are *real*. He cannot afford to abolish the reality of relations because without them he could not get one of the key building blocks of his metaphysical system – the substantial incompossibility that underlies the proliferation of mutually exclusive worlds as possible alternatives to one another.[8]

Nevertheless, a serious difficulty arises in this connection.

3. THE REDUCIBILITY OF RELATIONS

Leibniz espoused a metaphysical theory of "(real) existence" according to which only substances and their properties are real.[9] Other things – pre-eminently including *relations* between substances – are accordingly "things of the mind" (*entia rationis*), mental products belonging to the realm of phenomenal *appearance* rather than that of existential *reality*.[10]

> [A] relation, in this third way of considering it [e.g., as something beyond the modifications of the two relata], is indeed *outside* the subjects; but being neither a substance, nor an accident, it must be a merely ideal thing, the consideration of which is nevertheless useful.[11]

How can this insistence on the *ideality* of relations be reconciled with their indispensable role in underwriting incompossibility, a role which clearly requires their *reality*?

The answer is that relations – while from a certain point of view indeed "ideal" – nevertheless have a solid foothold in undoubted reality, in the modifications of substances. Indeed a relation only "exists" insofar as it roots in the properties of its relata. Relations do not have reality in their own right, but a *dependent* reality correlative with their inherence in the related terms. To see how this Leibnizian doctrine is to be understood, we must examine his important thesis of the *reducibility* of relations.[12]

Consider, to begin with, the relational thesis

Titius is wiser than Caius.

Its substance-descriptive foundation lies in the two predicative facts:

Caius is somewhat wise.
Titius is very wise.

To these we must conjoin certain "necessary truths", namely the general (definitionally guaranteed) facts:

(1) wiser = superior in point of wisdom
(2) "very" represents a degree superior to "somewhat".

The predicative facts from which we set out thus clearly suffice for extracting the substantival relations at issue (given universal truths).

Unfortunately, matters are not always so simple. Relations need not always boil down to *mere conjunctions* of predicative facts; more complex modes of compounding are sometimes necessary.[13]

Take the relational fact "Adam is the father of Cain". Leibniz maintains that this reduces to the fact that (1) Adam has two features:

(1) Being a father
(2) Being a father in virtue of (*propter*) Cain's being a son.

and (2) the fact that Cain has the two features:

(1) Being a son
(2) Being a son in virtue of Adam's being a father.[14]

The relation (relational fact) in question, namely "Adam is the father of Cain", thus issues from a series of predicational facts about the relata, predictational in which to be sure, a compounding operator – the reason-adducing "in virtue of" (*eo ipso* or *propter*) connective – plays a role. Sometimes Leibniz assigns this task to a "since" or "seeing that" or "insofar as" connective, specifically *quantenus*, or again, sometimes to a "thereby" (*eo ipso*) connective construed in the manner of "Adam is a father and thereby (*es ipso*) Cain is a son". All these are variations on the same reason-giving theme of a fundamentally syncategorematic *ratio-essendi* specifying connective conjoining strictly predicational facts.[15]

There are accordingly two sorts of relations: *Relations of comparison* (*relationes comparationis*) as with "Titius is wiser than Caius" and *relations of connection* (*relationes connexionis*) as with "Paris loves Helen".[16] In the former case, it suffices to "reduce" the relation if aRb can be extracted from strictly predicational information about the terms at issue, viz for $P_R a$ and $Q_R b$, for suitably nonrelational predicates P_R and Q_R, together with purely conceptual (logical and definitional) principles, i.e.,

$$aRb \equiv P_R a \;\&\; Q_R b$$

With connexive relations, however, the situation is a bit more complicated. Here too the relationship inheres in predicational information, but this time attributional facts are also needed:

$$aRb \equiv P_R a \;\&\; Q_R b \;\&\; (P_R a \;@\; Q_R b)$$

or more simply

$$aRb = P_R a \ \& \ (P_R a \ @ \ Q_R b).$$

Here @ stands for "is attributable to", i.e., represents the *eo ipso* or *propter* connective of "reason-why" attributability. In this way, whenever *aRb*, then the relationship at issue is always built into the individuating descriptions the terms being related; it is inherent in the fact that suitable compoundings of *Pa* and *Qb* obtains. To be sure, we here require not only conjunction but attributability. But the attributability linkage now at issue is something which, though itself clearly relational, involves – like conjunction itself – a relationship that obtains not between substances, but is a rational connection between purely predicational facts about substances.[17] It represents a relationship not between *substances* but between *facts*.

This reduction of relationships to the property-structure of the relata explains how Leibniz can say that the monads "correspond to each other through their own phenomena and not by any intercourse or connection".[18] Recourse to this grounding relationship shows how Leibniz proposes to resolve the difficulty that a doctrine of predicate-reducibility must at one and the same time ascribe relations to terms taken severally and singly and yet with reference to one another. As Simplicus had observed, things are related when they exhibit an "inclination towards one another".[19] It was Leibniz's insight and contention that all of the myraid modes of connective relation among substances can be reduced to one basic fundamental pattern in which only (nonrelational) predications figure: For any R there are predicates P_R and Q_R such that:

$$(\forall x)(\forall y)\ xRy \equiv [P_R x \ \& \ (P_R x \ @ \ Q_R y)]$$

What the *properties* of substances are is simply a matter of their predicationally descriptive makeup (their *identifying* notions), but the *reason-why linkages* that govern their interrelationships are a matter of the aims and purposes inherent in the will of God.[20]

This analysis indicates another basis for Leibniz's insistence that relations be construed as things of the mind – the fundamental *propter* grounding of reason-why attributability which serves as a rationale-presenting device is fundamentally mind-oriented. Most importantly, this basing of all relationships on one invariant linkage of reason-why grounding shows how literally Leibniz is prepared to assimilate the two basic senses of *ratio* as on the one hand meaning *reason or ground* and on the other hand meaning *relation or mode-of-connection.*

Thus, *substantial relations will always inhere in the nonrelational properties of the relata at issue* through the linking mediation of the purely propositional connections of conjunction and attributability.[21]

To be sure, this is not a matter of a *logical* reduction of a relation into something nonrelational: we are still left with something relational. Rather, it represents something relational within the combinations among purely predicative facts comprised in the notions definitive of the substances at issue. Relations of substantival connection are reduced to ones of rational grounding. In sum, reality is completely characterizable – as far as its *description* goes – in strictly predicational terms: aRb is always extractable from suitable compoundings of $P_R a$ and $Q_R b$. Leibniz, as is well known, devoted a good deal of energy to this "reducibility project" of showing how relational statements can be recast in terms of complexes of statements that take a non-relational form.[22]

When aRb is said to be "reducible" to predicational information about a and b, do these predicates not include *rational* predicates themselves? Consider Leibniz's own example

$$a \text{ is the lover of } b = a \text{ is a lover and } \textit{eo ipso } b \text{ is beloved}$$

"a is a lover" (i.e., a has the property $(\lambda x)(\exists y)xLy)$ and "b is beloved" (i.e., b has the property $(\lambda x)(\exists y)yLx)$ are clearly *relational* properties. We seemingly do not manage to eliminate relations at all. And so, when aRb is "reduced" to the combination of $P_R a$ & $(P_R a @ Q_R b)$ the question arises whether, given the aims of Leibniz's reducibility program, the properties at issue – P_R and Q_R – will themselves have to be genuinely qualitative (wholly nonrelational), or whether it will simply do to let P_R be "bearing R to something" or $(\lambda x)(\exists y)xRy$ and Q_R be "having something bear R to it" or $(\lambda x)(\exists y)yRx$. Now, as I see it, Leibniz has two options here. One is to simply take such *universalized* relational properties in stride, conceding that one-place predicates are relational in their meaning-content, but insisting that this is quite harmless because any such relationality *sub ratione generalilatis* involves no reference to any particular individuals. Alternatively, Leibniz could maintain that when a is a substance that has the generically relational property $(\lambda x)(\exists y)xRy$, there must be some complex of genuinely monadic properties $P^1{}_R, P^2{}_R, \ldots P^n{}_R$ such that, in the final analysis, $(\lambda y)aRy$ is *analytically equivalent* to $P^1{}_R a$ & $P^2{}_R a$ & ... &$P^n{}_R a$. Thus, for example, "being a father (to someone)" would have to be *equivalent* to a (merely conjunctive) complex of strictly intrinsic denominations: fatherhood must then be somehow correlative with some internal modification – some characteristic added furrow of the brow. The relational properties that a substance has *sub ratione generalitatis* must be somehow encompassed within its purely intrinsic denominations. So far as I can see, either alternative is open to Leibniz without any sacrifice in his other commitments.[23] And either one would achieve his *metaphysical* purpose of embedding the relations between substantival terms in the substance-abstractively descriptive characterization of the substances at issue. (It bears repeating: we are not concerned with the elimination of relations as a *logical* resource.) The key point, either way, is that the individual concept of a substance will be complete where all of its *nonrelational* properties are given in their suitable compounding by & and @.

Thus in the final analysis, Leibniz's position on the reducibility of intermonadic relations comes down to something fairly simple and straightforward:

(1) All such relations are reducible to combinations of merely predicative propositions as embodied in the complete individual notions of the substance at issue.

(2) In these combinations, no more than logical or quasi-logical connectives are invoked – above all conjunction (&) and reason-why-attributability (@).

The crucial point is that a proposition in which a relation to another substance is attributed to an individual can always be reconstructed in terms of complex propositions in which the only properties being ascribed to individuals are generally descriptive properties in which no reference to individual substances are involved. Where *intersubstantival* relations are concerned, a reduction to a grounding in descriptive predicates is always possible.

Accordingly, relational statements are true precisely when they are "well founded" in the properties of the substances at issue.[24] Different substances do not share the same modification:[25] paternity in David is *fatherhood of Solomon*, paternity in Adam is *fatherhood of Cain and Abel*, and these are emphatically distinct – they eventually unpack in very different ways when we go over to their predicational analyses. Intersubstantial relations are founded on the properties and property-ramifications of the substances that represent the relata at issue. There is nothing to a relation over and above what is embodied in the framework of such a foundation.

This grounding of its relations within the properties which (for Leibniz) define a thing as the very thing it is – through their role in its complete individual concept – leads us inexorably to the consequence that "a monad always expresses within itself its relations to all others".[26] We see, too, the ground for Leibniz's insistence on the *concreteness* of intermonadic relations and for his subscription (in the case of relations) to Avicenna's doctrine that it is altogether impossible "that numerically one and the same accident may exist in two substances".[27]

4. RELATIONAL REDUCIBILITY AND INCOMPOSSIBILITY

Let us now turn to how incompossibility operates. Suppose that *a*'s description is such that *aRb* obtains, or equivalently that

$$P_R a \; \& \; (P_R a \; @ \; Q_R b)$$

And suppose further, that *b* did not cooperate in this regard, i.e., that there is no property Q actually to be found in *b*'s individuating description such that $P_R a \; @ \; Q_R b$ obtains. Then *a* and *b* are incompatible substances. Then recourse to attributability at once affords us with machinery to assure the incompossibility of substances. At the outset we observed that Leibniz needs relations to constrain incompossibility. We now see that this theory of relational reducibility does indeed furnish an account of relations through which the principle that relations can provide for incompossibility is secured.

As we have seen, when *a* is related to *b* (i.e., when *aRb*), then its R-relationship to *b* is always built into the complete individual notion of *a*. But, of course, the defining notions of substances do not (or should not) make mention of other *particular* substances. And so when we "reduce" *aRb* as amounting from *a*'s standpoint to

$$P_R a \; \& \; (P_R a \; @ \; Q_R b)$$

we should in fact generalize *b* with respect to any and every substance x answering to *b*'s individuating description (D_b):

$$P_R a \; \& \; (\forall x)[D_b x \supset (P_R a \; @ \; Q_R x).$$

Since D_b is *b*'s identifying or individuating description, the condition that $D_b x$ is to obtain effectively limits the acceptable values of x to b alone, via the Principle of Identity of Indiscernibles.[28] But the harm is undone, because we now proceed *sub ratione generalilates*, without reference to other pre-identified substances.

But a further possible difficulty now arises. Let us suppose that aRb. Then the identifying description of a must encompass an effective equivalent of $(\lambda x)xRb$ or, more fully, $(\lambda x)(\forall y)[D_b y \supset xRy]$. And similarly, of course, the identifying description of b must encompass an effective equivalent of $(\lambda x)aRx$ or, more fully $(\lambda x)(\forall y)[D_a y \supset yRx]$. If this is to be workable, then the identifying description at issue (viz. D_a and D_b) must, of course, themselves be wholly free of particularized relational involvements. They must be wholly general, since otherwise the whole reductive process would self-destruct. (If a's identifying description involved its relations with b, and b's identifying description involved its relations with a, then we would be in difficulty, since in describing a we would require a finished (preexisting) description of b, while to describe b we would need a finished (preexisting) description of a. The way around this difficulty is to extricate from the identifying descriptions of individuals any reference to other identified individuals, and to insist that one must proceed here wholly in terms of nonrelational generalities.

It was said above that on Leibniz's theory of relations, if aRb, then this is something that can be deduced from an informational basis consisting of a's identifying description and b's identifying description. We can now give this thesis a somewhat more accurate formulation.

If aRb, then a's individuating description must embody the relational property

$$(\lambda x)xRb$$

or equivalently (given the Identity of Indiscernibles)

$$(\lambda x)(\forall y)[D_b y \supset xRy]$$

or more fully yet, given relational reducibility:

$$(\lambda x)(\forall y)[D_b y \supset [P_R x \ \&\ (P_R x \ @\ Q_R y)].$$

And of course, if *a* has *this* property then it follows, not (to be sure) categorically, but at any rate hypothetically, that *a* stands in the relation *R* to any substance answering to *b*'s identifying description. Moreover, *a* would clearly be incompatible – i.e., *logically* incompatible – with any substance that otherwise answers to *b*'s description but lacks the relationship needed to underwrite *b*'s possession of $(\lambda x)aRx$ (in the counterpart to the fuller version obtainable along the preceding lines).[29]

5. REDUCIBILITY NOT A LOGICAL BUT A METAPHYSICAL THESIS

It must be stressed that the reducibility of relations is, for Leibniz, a *metaphysical* rather than a *logical* thesis. It does not obtain for relations between any two terms whatsoever, but specifically for relations between *substances*. Leibniz does not – and need not – say that the relational fact that five is greater than three is reducible to the (non-relational) properties of these numbers. But with *substances*, the case is otherwise. God conceives of a substance wholly as it is – with complete information – and so if that substance is related to another, God conceives of it as so related. For this relationship is built into the very notion that is definitive of the substance as the substance it is: since this notion is complete, it incorporates the relationship at issue.

Given their reducibility, the relations of a particular substance can be no more *extrinsic* to its nature than can its properties or denominations; and

> There are no purely extrinsic denominations which have no basis at all in the denominated thing itself. For the concept of the denominated subject necessarily involves the concept of the predicate.[30]

Even as predicate-containment is the key to the truth of property-attributions, so its predicative structure (attributability connections included) is the key to the truth of relational propositions regarding a substance. A relation is an accident, and

> Accidents cannot be detached from substances and march about outside of substances, as the sensible species or the Scholastics once did.[31]

This reducibility of relations is a matter of how *substances* are conceptualized, and turns crucially on the completeness of conceptualization. Every relation between substances has a *fundamentum in re* through its embedding in the descriptive makeup of the things at issue.[32] This thesis that relations are embedded in and extractable from the description-internal properties of their relata, as built into the completable individual notions thereof, is a characteristic and central doctrine of Leibniz's metaphysics.

Leibniz is concerned to establish not the logical eliminability of relations but their metaphysical dispensability at the level of individual substances. It needs to be stressed here that what is at issue is not a strictly *logical* doctrine, but a *metaphysical* one. The theory turns on the identification of substances through complete individual notions and the operation of a grounding relationship that is able to build rational connections into the defining conceptions of substances (thus making their defining descriptions more than a matter of merely *conjoining* predications, though not more than a matter of *connecting* them).[33]

A substantival relation is thus always concept-internalized and not advantitious, that is, added *ab extra*. Its relations are included within the concept of a thing and nowise independent of it: the defining notions of substances embody grappling hooks into the environing world. Indeed, a world is *self-inconsistent* if one of its substances has a relation in its concept whose relatum does not exist in that world.

This way of accommodating relations thus at once resolves the problems of substantival incompossibility. At one time Leibniz had written:

> However until now it has not been known to people whence the incompossibility of different things arises, or how it is possible for different essences to conflict with one another, since all purely positive terms seem to be compatible with one another.[34]

But, of course, this problem is resolved once reason-why linkages are superimposed upon the descriptive notions of substances. If the Cain of a world is a brother by virtue of the presence in that world of Abel (or, rather, of a substance of Abel's predicative description) while the Abel of a world is an only child, then it is clear how these are incompossible – i.e., could not be co-realized within one particular world.

These considerations bring to the fore the importance of relational reducibility. Leibnizian monads are windowless. They admit no *influxus physicus*, no readjustment, no reaccommodation to the changes of others. On the other hand they agree, standing in a delicately attuned mutual accommodation, and exhibiting an all-pervasive coordination of mutual *expression* – i.e., interrelation.[35] This mutual accommodation of the substances of one selfsame world is built into their concepts; their relationships are embedded within their complete individual notions. As Leibniz wrote to Arnauld:

> I say that the concept of an individual substance contains all its changes and all its denominations, even those commonly called extrinsic, that is to say which pertain to it only by virtue of the general interconnection of things, and because it expresses the whole universe in its own way. For it is always necessary that there be some foundation for the connection of the terms of a [true] proposition which must be present in their concepts.[36]

The key fact is not that substances do not have relations, but that these are *and have to be* impressed with the inner design of the propositional structure of their defining descriptions. This is how it comes about that every simple substance has relations which express all the others [i.e., substances], and that consequently it is a perpetual living mirror of the universe.[37]

The idea that (as Leibniz repeatedly says) each substance *mirrors* the whole universe from its own point of view is meant to suggest that it itself bears within its own qualitative make-up the imprint of the nature of its fellows.

It is important to recognize, however, that even if such (partially described) substances are *pairwise* compossible, they might still prove to be incompossible *in toto*. For suppose that *a*, *b*, and *c* are such that

(1) *a* has (*inter alia*) the feature of "being unique in lacking *P*"[38]

(2) *b* has (*inter alia*) the feature of "being unique in lacking *Q*"

(3) *c* has (*inter alia*) the feature of "being unique in point of having a having a lack/possession status for *P* different from that it has for *Q*".

Here any two of these individuals will be mutually compossible. But when all three are put together, their conjoint realization in a single world will clearly be infeasible. However, when we shift from *partially* described substances to complete individual notions this sort of situation does not arise. Substantival compossibility now becomes transitive. For here we have a "block universe" where everything is inseparably connected with everything else. When Adam sins, all mankind comes to reflect his sinfulness.

Their ability to constitute a possible world – their *synoptic* compossibility – is a global and comprehensively systematic feature of a group of possible individuals. Each substance within a possible world carries within itself an ineradicable imprint of all the rest:

That each singular substance expresses the whole universe in its own way, and that in its concept are included all of the experiences belonging to it together with all of their circumstances and the entire sequence of exterior events.[39]

None of its substances can be abstracted from that world and none adjoined to it without undoing the intricately woven fabric of compossibility relationships.

The pervasiveness of the connection of rationale-provision inherent in the Principle of Sufficient Reason means that every facet of a substance ultimately has relational involvements:

[T]here is no term so absolute or detached as not to included relations and the perfect analysis of which does not lead to other things and even to all others; so you can say that relative terms indicate *expressly* the relations they contain [while nonrelative terms merely do so *implicitly*].[40]

The pervasiveness and the reality of relations are reflected in the fundamental Leibnizian principle *tout est liè* ("Everything is interconnected").

[I]n metaphysical rigor, there is no such thing as a purely extrinsic denomination (*denominatio pure extrinseca*) because of the real connection of all things.[41]

The pre-established harmony assures that the actual world is a seamless causal system. The reducibility of relations assures that every possible world is a seamless conceptual system.

6. THE REALITY OF INTERMONADIC RELATIONS

The fact that all intersubstantival relations are reducible to complexes of predications means that an intermonadic relation has no independent existence of its own, over and above that of the related substances and their (genuinely nonrelational) properties. Their complete individual notions afford the strictly predicational information about the substances of a possible world that is always sufficient to yield by derivation all the facts about their relationships as well. A relation is nowise a *tertium quid* existing on its own, independently of the relata. It is a syncategorematic compound of predicational facts, but not something further, with an independent status of its own. Relations have no standing apart from the existence of the relata and their properties.

Construed as a connecting linkage that is distinct from and stands outside of the items it links, a relation lies wholly in the mind of the beholder. If one insists on viewing a relation as a connection existing somehow intermediate between the two extremes and apart from them – straddling them, as it were – then, Leibniz holds, there is not such entity to be found in nature:

> You will not, I believe, admit an accident that is in two subjects at once. Thus I hold, as regards relations, that paternity in David is one thing, and sonship in Solomon another, but the relation common to both is a merely mental things of which the modifications of singulars are the foundation.[42]

But such remarks go no further than to show that relations must not be thought of in this term-detached way. They must not be construed as maintaining that relations do not exist or are somehow unreal. It is not that substances do not have relations – *au contraire*, this is one of their most crucial aspects. A substance bears relation to everything else in its world (*Sympnoia panta*).

Accordingly, Leibniz does not hold a relation to be something unreal – something fictional or illusory or the like. Relations are (or can be) perfectly *real*: they are phenomena and indeed well-founded phenomena. But though real, relations are not *real things*. For they are not literal things at all – not *substances*. In the system of Leibniz, to be a phenomenon is a to be real without being a *thing*. (Only substances are real things and relations are not substantial.) Relations are thus real because they exist in and through the characteristics of real things, being embeddable in the make-up of substances. Yet relations are secondary and supervenient: they arise from and inhere – lock, stock, and barrel – in the properties constituting the identifying description of things. Still, this clearly gives them a perfectly workable *fundamentum*. Since its relations are always built into the complete individual notion that is identificatively definitive of a particular substance, there is no way of tampering with the relations of a substance: Its concrete relations are always hypothetically necessary features of it as the very substance it is. (Of course, for Leibniz this does not stop them from being contingent.)

We often hear it maintained that Leibniz does and must deny the *reality* of intermonadic relations. Thus Julius Weinberg writes:

If there were real connections among things, the doctrine that the predicate of every true affirmative proposition is contained in the subject could not be maintained, and this doctrine is surely central in Leibniz's thought.[43]

But the objection that is at issue here does not survive critical scrutiny. It is not the *reality* of relations that is incompatible with Leibniz's theory of truth but their *independence*. The aim of his theory of predicational imbeddability of intermonadic relations is precisely to show how relations are compatible with the *praedicatum inest subjecto* approach of Leibniz's concept-containment theory of truth. What is required, however, is that this reality of relations not be an *independent* reality, that relations not be a *tertium quid* existing independently of substances and their properties, in sum, that relations not be purely and wholly extrinsic denominations, but always have a basis in the modifications of singulars.

To be sure, relations taken as *purely* relational – i.e., as connecting linkages standing between, and thus apart from substances they relate – are mental.[44] But even from this perspective, intermonadic relations are not *merely* mental. They are not something *wholly* imaginary or and correlative, like a mirage or a delusion.

In virtue of the fact that substantival relations are always property-reducible, they have a firm foothold in reality.

Moreover, that master mind, the mind of God, has a solid grasp on relational facts. And so it transpires that relations, even when viewed as *a tertium quid* standing between their relata, have a certain quasi-reality that accrues to them in view of their presence to the mind of God.[45] It must not be thought that Leibniz is a radical subjectivist regarding relations – one who holds that they exist only "in the mind of the beholder". He emphatically does *not* hold this; indeed not even when the beholder is God. For when God recognizes a relational fact, his thought does not thereby *create* the relation at issue, but acknowledges something independently actual and conceptually prior:[46]

> Yet, although relations are of the understanding, they are not groundless or unreal. For the divine understanding is the origin of things; and even the reality of all things, simple substances excepted, exists only on the basis of the perception of the phenomena of simple substances.[47]

In relating two substantive terms, the mind of God reflects a perfectly objective basis nowise created by it in this act of cognition – a basis within the conceptual order of things and their reasons for being.

Leibniz, then, does not teach that relational statements are meaningless, or that intersubstantival relations do not exist, but that all relations that can obtain between substances must inhere in their predicates. Relations are "the work of the mind" (in Russell's phrase) because, when substances are under discussion, relational statements are never "ultimate facts", they are always based upon a suitable *fundamentum in rebus* as derivative consequences of complexes of predications. For Leibniz, relational statements about substances can never afford information about them that is not given more fully and adequately by suitably combined predictions. But such reducibility is nowise at odds with the reality of relations.

In his *Commentary on Aristotle's Categories*, Simplicus objected to the Stoic claim that if we deny objective reality to the connections between things, we thereby abolish the union and harmony of nature.[48] Geometry, music, and all sciences of order would be otiose if the connections among things to which they address themselves themselves did not exist. Leibniz is also deeply committed to the logic of this argument. His thesis that relations have no existence *independently* of substances must be so construed as to avoid denying reality to relations, and to refrain from insisting they *merely* lie in the eyes of beholder. His theory of the reducibility of relations, of their well foundedness through embeddability in the properties of substances, is thus a crucial aspect of his metaphysics in enabling him to maintain the *reality* of relations in the face of the other fundamental commitments of his system which have it that only substances and their properties are real.

7. ABSTRACT RELATIONS

It should be stressed that our discussion has throughout focused on *concrete* relationships between substances, and not their *abstract* relations as such – for example, placement in some generic framework of positions. It goes without saying that for Leibniz such abstractions are *entia rationis* that have no existential status save the mental.[49] And even here they are of an inferior condition. For they are not only mental or mind-correlative but mind correlative only for our imperfect minds.[50] In thinking about substances. God has no need for abstractions. For abstractions arise when the details are omitted, and an omniscient God does not omit the details. At the divine level, nominalism holds good – the eye of God's mind grasps in a *totum simul* glance the vast myriad of details that our imperfect minds only comprehend incompletely. Abstract and therefore inherently incomplete conceptions (with "red" = "looking somewhat like this") have no role in the thought of God. Abstract relations are the cognitive instruments of creatures who must *faute de mieux* think confusedly even as they perceive confusedly. And so for Leibniz, abstract relations are indeed something *purement ideale* – they lack that reality which their presence to the mind of God invariably imparts to concrete substantival relationships. Moreover, they lack that *fundamentum* in the domain of real things which always characterizes *intersubstantival* relations owing to their inherence in the descriptive constitution of their relata. For, as we have seen, this reducibility of relations is not a *logical* fact concerning relations in general, but a metaphysical fact specifically about *substances*.

NOTES

[1] Benedictus Spinoza, *Short Treatise*, Pt. I, Chap. 10.

[2] Pierre Gassendi, *Syntagmata Philosophiae Epicurei*, Bk. II, Sect. 1, para 15.

[3] Thomas Hobbes, *Elements of Philosophy*, Pt. II, Chap. xi, §6; *English Works*, ed. by W. Molesworth (London, 1839–45), Vol. I, p. 135.

[4] Bertrand Russell, *A Critical Exposition of the Philosophy of Leibniz* (London, 1900; 2nd edn. 1937), p. 13. Again: "After he [Leibniz] has seemed for a moment to realize that relation is something distinct from and independent of subject and accident, he thrusts aside the awkward discovery" (Ibid.).

[5] "And the perceptions or expressions of external things come into the soul at the proper time by virtue of its own laws, as in a world apart, and as if nothing existed but God and itself". (*New System*, §14; Loemker, p. 457). On Leibnizian principles, this holds not only of *souls* (=conscious substances), but of substances in general.

[6] The point is not that a Leibnizian substance must be "free from all outer relations" (as Kant puts it at *CPuR A*274=*B*330), but that it must internalize any such relation in its own concept.

[7] Couturat, *Opuscules*, p. 530; Loemker, p. 169 and compare p. 263.

[8] This problem was raised by Bertrand Russell in *A Critical Exposition of the Philosophy of Leibniz* (London, 1900, 2nd edn., 1937), p. 67. It was emphasized in the middle 1960s in the author's *The Philosophy of Leibniz* (Englewood Cliffs, 1967). The issue was picked up by Benson Mates in his paper on "Leibniz on Possible Worlds" in B. van Rootselaar and J. F. Staal

(eds.), *Logic, Methodology and Philosophy of Science III, Proceedings of the 1967 International Congress* (Amsterdam, 1968), pp. 507–529 (esp. pp. 519–521). (This essay was reprinted in H. Frankfurt (ed.), *Leibniz: A Collection of Critical Essays* (New York, 1972), pp. 355–364). See also Jaakko Hintikka, "Leibniz on Plenitude, Relations, and the Reign of Law", *Ajatus* Vol. 31 (1969), and Hide Ishiguro, "Leibniz's Theory of the Ideality of Relations", ibid., pp. 191–213; both reprinted in the Frankfurt anthology.

[9] This is essentially the doctrine of Avicenna (and the Arabic Aristotelians in general) who held that everything real is either substance or attribute. These teachings reached Leibniz via the schoolmen (especially Aquinas and Scotus). Relations cannot have an *independent* existential status in Leibniz since a relation involves (at least) two relata whereas *ens et unum convertuntur* and "What is not truly one entity, is not truly an entity". (Letter to Arnauld of 30 April 1687: in *Lettres de Leibniz à Arnauld*, ed. by G. Lewis (Paris, 1952), p. 69.)

[10] Leibniz's theory is at many points identical with that of those proto-nominalists, the ancient Stoics. The Stoics taught that only individual bodies exist, and rejected forms and universals. Only *substances* and their *qualities* have independent reality: the remaining categories ("state" [*pōs echon*] and "relation" [*pros ti*]) embrace the "incorporeals" which are characterized as "the meanings of spoken words" (*ta lekta*) and thus have no extramental reality. Accordingly, relationships are subjective and have no mind-independent ontological status as objective features of the world in their own right. (See Emile Bréhier, *Chrysippe et l'ancien stoicisme* (Paris, 1910; 1951), p. 70 and *La Théorie des incorporels dans l'ancien stoicisme* (Paris, 1913), p. 20.)

[11] Fifth letter to Clarke, §47; Loemker, p. 704.

[12] On this topic see especially G. H. R. Parkinson, *Logic and Rationality in Leibniz's Metaphysics* (Oxford, 1965), pp. 45–55.

[13] Leibniz's recognition of this fact inheres in his distinction between "relations of comparison" and "relations of connection":

> Relationes sunt vel comparationis vel connexionis. Relatio comparationis ex eo nascitur inter A et B, quod A reperitur in aliqua propositione, et B in alia propositione, Relatio connexionis ex eo quod tam A et B sunt in una eademque propositione (quae in relationem comparationis resolvi non potest[)]. Nam alioqui etiam relatio comparationis foret relatio connexionis, nam formari potest una propositio comprehendens A et B, nempe A est similis B. Sed ea resolvitur tandem in duas quarum una singulatim agit de B, altera sigillatim [sic.] de A, verbi gr. A est ruber et B est ruber, et ideo A est similis (quoad hoc) ipsi B. per A autem et B intelliguntur res seu individua, non termini. Sed quid de his dicemus: A existit hodie et B etiam existit hodie, seu A et B existunt simul? An erit haec

relatio comparationis, an connexionis idem est de coexistentia in eodem loco. (MSS [Hannover] LH, IV, vii., C, fol. 17r. Quoted in Massimo Mugnai, "Bemerkungen zu Leibniz' Theorie der Relationen", *Studia Leibnitiana*, Vol. 10 (1973), pp. 2–21; see p. 20.)

[14] Note that we must here construe such relational processes as "fathering a child" and "being fathered" as predicative characteristics of an individual.

[15] What is at issue here, however, is not a *relation between things* but a *grounding relationship between facts*. For further details regarding Leibniz's approach see G. H. R. Parkinson, *Logic and Reality in Leibniz's Metaphysics* (Oxford, 1965), pp. 39–52. Compare also Benson Mates, "Leibniz on Possible Worlds", in B. van Rootselaar and J. F. Staal (eds.), *Logic, Methodology and Philosophy of Science III, Proceedings of the 1967 International Congress* (Amsterdam, 1968), pp. 507–529 (esp. pp. 519–521), and Jaakko Hintikka, "Leibniz on Plenitude, Relations, and the Reign of Law", op. cit.

[16] Cf. Note 13 above.

[17] The linkage of reason-why attributability that is operative here is still in a sense relational. The machinery of truth-functional logic does not enable dyadic relations to be reduced to monadic predicates alone. (Cf. C. I. Lewis and C. H. Langford, *Symbolic Logic* (New York and London, 1932), pp. 387–388.) But what we have here is not a relationship between *substances*, but a relationship between purely predicational *propositions*. And it is a uniform relationship that does not differ from case to case, but (like conjunction itself) is syncategorematically invariant across the entire board. The crux is that all of the thing-characterizing attributions at issue are purely predicational – in that the P or Q at issue are nowise relational. We thus have an authentic reduction to genuinely predicational statements.

In discussing my earlier exposition of Leibniz's views on relations, Jaakko Hintikka wrote:

> Nicholas Rescher ascribes to Leibniz the view that all relations among individual substances are reducible to and derivable from the predicates of their respective substances . . . Rescher presupposes that all those predicates to which relations are reduced are monadic in the sense of not even containing implicitly relational components. This means overlooking . . . [the possibility] that relational statements can be reduced to statements in each of which a complex predicate is ascribed to one and only one of the relata. These complex predicates may still involve relational concepts . . . (Leibniz on Plenetude, Relations, and the Reign of Law", in *Ajatus*, Vol. 31 [1969]; see pp. 9–10 of the separatim reprint.)

This view of a so-called "reduction" to predicates that may themselves still involve relations seems to me untenable. I adhere to my earlier view that whatever predications are operative at the post-reduction stage must be "monadic in the sense of not even containing implicity relational components". However, my earlier view of reducibility is indeed open to a charge of oversimplicication. For it envisaged only one syncategorematic connective at work within the purely monadic predications of the post-reduction statement, namely conjunction (&), whereas there are in fact two: conjunction and attributability (@).

18 Letter to des Bosses of 26 May 1712; Loemker, p. 602. This shows how mirroring, harmony and all the other concepts of reciprocal coordination among monads must be construed – viz. in terms of the due accommodation of their defining individual notions.

19 In *Aristotelis Categorias Commentarium*, ed. by C. Kalbfleisch (Berlin, 1907) p. 175.

20 ... *chaque substance individuelle exprimant tout l'univers, dont elle est partie selon un certain rapport, par la connexion qu'il y a de toutes choses à cause de la liaison des resolutions ou desseins de Dieu.* (GP, Vol. II, p. 51.)

21 [*Relationes*] *ipsae per se nullam denominationem intrinsicam constituant adeoque esse relationes tantum quae indigeant fundamento sumto ex praedicamento qualitatis seu denominatione intrinseca accidentali.* (Couturat, *Opuscules*, p. 9.)

22 Cf. for example the opuscule on "Grammatical Thoughts" published in abbreviated form in G. H. R. Parkinson (ed.), *Leibniz: Logical Papers* (Oxford, 1966), pp. 12–15.

23 But see Hidé Ishiguro, "Leibniz's Theory of the Ideality of Relations", in H. Frankfurt (ed.), *Leibniz: A Collection of Essays* (op. cit., pp. 191–214 (see especially pp. 207–210)). Ishiguro quite properly stresses that a property like $(\lambda x)(\exists y)xRy$ – "having R to something" – is inherently relational in its meaning-content. Thus, in resorting to it, one has not eliminated relationality in its *logical* aspect. On the other hand, since any reference to identified individuals is now eliminated, all the needs of a metaphysical reduction-program are met.

24 We have been dealing throughout this discussion with relations between *substances* (monadic relations). Relations among (well founded) phenomena are, of course, also held to be reducible by Leibniz, doubly so because they reduce to inter-monadic relations which then, in turn, come to be grounded in the properties of the several substances involved. Some interesting observations on phenomenal relations occur in *Nouv. Ess.*, Bk. II, Chap. ii, §§4ff.

25 *Neque enim admittes credo accidens, quod simul sit in duobus subjectis.* (GP, Vol. II, p. 486.)

26 GP, Vol. II, 457; Loemker, p. 605. *Hinc omnes propositiones quas ingreditur existentia et tempus, eas ingreditur eo ipso tota series rerum, neque*

enim τὸ nunc vel hic nisi relatione ad caetera intelligi potest. (Couturat, *Opuscules*, p. 19.)

27 Avicenna, *Metaphysica sive Prima Philosophia* (Venice, 1495; reprinted Louvain, 1961), Tract iii, Chap. 10. Leibniz repeatedly stresses this principle, for example at GP, Vol. II, pp. 457 and 586; 5th letter to Clarke, §42; *New Essays* Bk. II, Chap. xii, §7; etc.

28 The implicit universality of such a quantificational proposition indicates yet another way in which a substance "reflects the whole world". (Compare Jaakko Hintikka, op. cit., §§7–8.) It avoids the circumstance that *a*'s description need refer to any *particular* individual substance distinct from itself, its notion formed completely *sub ratione generalitatis.*

29 This analysis seems to me to resolve the issue posed in an interesting paper by F. B. D'Agostino as to whether substance-incompossibility does or does not rest on a strictly logical basis. See this author's "Leibniz on Compossibility and Relational Predicates", *The Philosophical Quarterly*, Vol. 26 (1976), pp. 125–138.

30 Loemker, p. 268.

31 *Monadol.*, §7. But cf. Loemker, p. 605, where we learn that despite its windowlessness, "a monad always expresses within itself its relations to all others". Leibniz insists time and again that while substances do have relations, these are built into (the individuating notions of) these substances themselves: "there must be a plurality of affections *and of relations* in the simple substance, even though it has no parts" (*Monadol.*, §13).

32 *[N] ulla datur denominatio adeo extrinseca ut non habeat intrinsecam pro fundamento, quod ipsum quoque mihi est inter κυρίας δόξας.* (To de Volder, Apr. 1702; GP, Vol. II, p. 240.)

33 This doctrine of the reducibility of relations to nonrelational monadic predicates is maintained to be Leibnizian orthodoxy in G. H. R. Parkinson, *Logic and Reality in Leibniz's Metaphysics* (Oxford, 1965) and in the writer's *The Philosophy of Leibniz* (Englewood Cliffs, 1967). It was subsequently called into question by Hidé Ishiguro in "Leibniz's Theory of the Ideality of Relations", in H. Frankfurt (ed.), *Leibniz* (Garden City, 1972) and *Leibniz's Philosophy of Logic and Language* (Ithaca, 1972). Ishiguro argues (quite rightly) that a scrutiny of Leibniz's proposals for the logical recasting of relational sentences shows that

> In most cases the sentences into which the original sentence is rewritten still describe relational facts, and the rewriting projects do not form any part of a general project of reduction in which nonrelational attributes are attributed to the relata. (Op. cit., p. 29.)

But this perfectly true observation overlooks the consideration that what is at issue with respect to intersubstantival relations is not a "general project" of *logical* reduction, but a specifically *metaphysical* project of showing how the relations among substances can be embedded in (and thus reduced to) their complete individual notions.

34 GP, Vol. VII, p. 195.

35 *Une chose exprime une autre* (*dans mon langage*) *lorsequ'il y a un rapport constant et reglé entre ce qui se peut dire de l'une et de l'autre* (GP, Vol, II, p. 112). All of the coexistent substances of the universe express one another.

36 GP, Vol. II, p. 56.

37 *Monadol.*, §56; The Russelian Leibniz who denies the reality of relations is a mere fiction.

38 That is, *a* has the relational property that "apart from myself, no individual in my environing world has not-*P*".

39 *Discourse on Metaphysics,* §9; Loemker, p. 308.

40 Nouv. Ess., Bk. II, Chap. 25, §10 (bracketed material added).

41 *Nouv. Ess.*, Bk. II, Chap. 25, §5. *Seposita tamen hac rerum universali sympathia, pro extrinsecis denominationibus haberi possunt.* (Quoted in Massimo Muguai, "Bemerkungen zu Leibniz's Theorie der Relationen", *Studia Leibnitiana*, Vol. 10 (1978). pp. 2–21 (see p. 15).)

42 GP, Vol. II, p. 486; Loemker, p. 609. Hidé Ishiguro observes (in *Leibniz: A Collection of Critical Essays* [op. cit., p. 200]) that this passage does not support the reducibility-of-relations construction "since neither paternity [i.e., "fathering someone"] nor filiation [i.e., "being fathered by someone"] are non-relational properties in any straightforward sense". This is true enough. But, of course, Leibniz's point is just that there are (complexes of) straightforwardly nonrelational properties in which both paternity and filiation can be embedded: the [genuinely nonrelational] "modifications of singulars" in which such relationships have their *fundamentum in rebus.*

43 Julius Weinberg, *Abstraction, Relation, and Induction* (Madison and Milwaukee, 1965), p. 114.

44 *Cette division des objets de nos pensées en substances, modes, et relations, est assez à mon gré. Je crois que les qualités ne sont que des modifications des substances et l'entendement y adjoute les relations.* (A. Vol. VI, vi, p. 145; quoted in Mugnai, op. cit., p. 5.)

45 *Porro Deus non tantum singulas monades et cujuscunque Monadis Modificationes spectat, sed etiam earum relationes, et in hoc consistit relationum ac veritatum realitas.* (GP, Vol. II, p. 438; cf. *Nouv. Ess.*, Bk. II, Chap. 30, §4.) "Cependant quoyque les relations soyent de l'entendement, elles ne sont pas sans fondement et realité. Car le premier entendement est l'origine des choses . . ." (A, Vol. VI, vi, p. 145 as quoted in Mugnai, op. cit., p. 5). Plotinus had already objected to the Stoic doctrine of relations that a theory which sees relations as mind-dependent *entia rationis* comes to grief on the fact that everyone realizes that there are relations he himself does not know about,

and thus presumably some relations no one thinks of. (*Enneads*, VI, 1, §7.) Recourse to the mind of God removes this difficulty.

[46] If God created relations, the incompossibility of world could not serve as a constraint upon his creative capacity and Leibniz's contention that God is *forced to choose* between alternative worlds would be lost – and with it many of the key doctrines of his whole system.

[47] *Nouv. Ess.*, Bk. II, Chap. xii. §5.

[48] Simplicius, *In Categorias*, Chap. vii; ed. by Kalbfleisch, p. 169. This view is restated by Aquinas, who held: *oportet in ipsis rebus ordinem quendam esse, hic autem ordo relatio quaedam est*. It is re-emphasized by Duns Scotus, who insisted that *relationes negare est negare totum ordinem universi*. See Julius R. Weinberg, *Abstraction, Relation, and Induction* (Madison and Milwaukee, 1965), p. 102. (Various data for the present discussion have been derived from this erudite and interesting book.)

[49] Such abstract relations arise when "l'esprit, non content de la convenance, cherche une identité, une chose qui soit veritablement la même, et la conçoit comme hors de ces sujets". (GP, Vol. VII, p. 401.) Such an abstract relation is "une chose purement ideale". (Ibid.)

[50] See §47 of Leibniz's fifth letter to Clarke.

ESSAY VI

LEIBNIZ AND THE PLURALITY OF SPACE-TIME FRAMEWORKS

1. THE QUESTION OF DISTINCT FRAMEWORKS

Leibniz advocated a theory of space (and time) as "relative" – that is, as relative to the physical things ordinarily said to be located *within* space (and time). He opposed the doctrine of Newton's *Principia* which cast space and time in the role of containers existing on their own and having a make-up that is indifferent to the things emplaced in them. Owing to the general tenor of his theory, Leibniz is sometimes seen as a precursor of Einstein and modern relativity theory. But this view is mistaken or, at any rate, misleading. For Leibniz – unlike Einstein and the modern relativists – is not thinking of the relativity of dynamical principles to the choice of a coordinate system *within* nature, so that we are involved in a situation of comparison from the perspective of various world-included frameworks. Rather, Leibniz's thesis that "space is relative to the things in it" has regard to the perspective of various alternative possible worlds taken as a whole. The mutual attunement of whatever is included in a common world is the foundation for space and time, which have no existence apart from the concordance of the mutual "perceptions" of substances (in Leibniz's sense of this term). "[T] here is no spatial or absolute nearness or distance between monads. And to say that they are compressed in a point or disseminated in space is to use of certain fictions of our mind when we seek to visualize imaginatively what can only be understood."[1]

As Leibniz saw it, the Newtonian theory of "absolute" space envisages this space as an entity in its own right, a content-indifferent container which would be filled up with different substantial content in the case of different possible worlds. His own theory that space is itself something content-relative implies – by way of contrast – that every possible world must have its own characteristic spatial structure. The issue comes down to a *metaphysical* – rather than *physical* – bone of contention. For in physics we study *this* world alone, whereas the point at issue is that of the question: Do *different* "possible worlds" have their own spatial structure or should they be conceptualized as different ways of filling up one single common content-indifferent space-time container?

2. SPATIALITY: THE CONCEPTION OF SPACE AS EVERYWHERE THE SAME

To begin with, we have to recognize that the *idea or conception* of space must (for Leibniz) be uniformly one and the same with respect to all possible worlds. This is true for space as it is for any and every *concept*. A possible world may or may not contain *men*, and its intelligent creatures may be very different from ours, but it cannot alter what *humanity* is. (The concept of humanity may not find *application* in some other possible world, but it cannot undergo *alteration* there.) In every world-setting space answers to the same conception – it is "the order of *coexistence*" (not – be it noted – the order of coexistents, which, after all, will differ from world to world). For Leibniz, every concept is what it is with respect to any and every possible world – the concept of space included. The *concept* of spatiality is world-uniform because it is world-indifferent. In this regard these concepts of space and time operate in exactly the same fashion as any concept whatsoever.

Let us, however, look at the matter from another point of view.

3. ONE WORLD, ONE SPACE

The ancient atomists had an interesting theory of possibility. Confronted with a question like "Why do horses not have horns, as cows do", they responded: "The hornlessness of horses is just a local idiosyncracy of *our* world – our own environing particular neighborhood in the universe. Somewhere else in the infinite vastness of space, there is another world, otherwise just like ours, in which horses *do* have horns." The atomists thus envisaged space as one vast framework in which all possibilities are concurrently encompassed.

Did Leibniz hold a view of this nature? Was space for him one vast, all-comprehending matrix that embraced the actual and possible alike – a *superspace* embracing all possible worlds along with our own, *actual* world?

Surely not. For Leibniz, every world has its own space. There is no superspace in which distinct possible worlds are co-located with one another. Leibniz, as we may say, was a "one-world, one-space" theorist.

A space for Leibniz is an order of *coexisting* substances, and distinct individuals in distinct worlds do not coexist with one another. (*Coexisting* substances are *a fortiori* com*possible*.) There are as many such orders as there are families of *compossibilia*. The limits of a space are coordinate with the realm of the substances comprising its correlative world.

Space for Leibniz is the order of coexistence (*ordo coexistentiae*). If Leibniz had defined space as the order of possible existents at large – rather than as the order of possible *co*-existents – then, to be sure, there would only be one single, all-comprehending space. For it is clear that different substances in different possible worlds do bear various relations to one another – the relation of *difference* for one thing, but also *similarity* (in various regards) and so on. But while there are cross-world *relations* among possible substances, there are not (as I interpret Leibniz) any cross-world *spatial relations*. Space is the order of *co*existence, and spatial relations are confined to coexistents. Distinct worlds are spatially disjoint – or better (since disjointness is itself a spatial term) they are spatially *unrelated* – somewhat like the dream-worlds of different people, (which is Leibniz's own illustration, as we shall shortly see).

4. DISTINCT WORLDS MUST HAVE DISTINCT SPACES

It is tempting to ask: "How can one Leibnizian substance possibly contradict another if they are located in different spaces?" But to ask this is to pick up the wrong end of the stick – it is to continue to be caught in the trap of the container theory. Substances are located in different spaces *because* they contradict one another: The world in which people otherwise like my parents had a daughter instead of a son for their only child has to be a different world from this one thanks to compatibility-considerations, and has to have its own distinctive spatiotemporal structure on the basis of these differences.

In the Paris period Leibniz enunciated a position which – as I interpret him – he continued to hold throughout his life:

> [T]here could exist an infinity of other spaces and worlds entirely different [from ours]. They would have no distance from us [nor other special relations to us] if the minds inhabiting them had sensations not related to ours. Exactly as the world and the space of dreams differ from our waking world, there could even be in such a world quite different laws of motion.[2]

Leibniz thus holds that every possible world has its own space as it has its own laws. There are many spaces, even as there are many law-manifolds. To say this does not countervail against the undoubted fact that what a *space* is like what a *law* is, is something that is uniform throughout all possible worlds. The concept for (or genus) is uniform even though its exemplifications (or instances) are distinct.

If one confronts the thesis that, for Leibniz, "space is one and the same everywhere, for all possible worlds", one must accordingly recognize that this is so in one sense and not so in another. It is so if we take in view the *concept* of space, but false if we take in view the *thing* to which this concept applies. For while space is – everywhere – the "order of coexistence", it turns out that what this order is is necessarily different in different worlds, since different worlds contain different (and incompatible) substances and these substances *internalize* such differences. (A difference in substance entails a difference in their relations which entails a difference in ordering relations.)

The space of the physical world, so Leibniz writes to Samuel Clarke, is not separable from its matter.[3] But space – spatiality as an order of coexistence – pertains not only to the actual world, but to every possible one. Yet in no possible worlds can space be separated from the substances that "fill" it. Space is nowise a content-neutral container.

For Leibnizian possible worlds, then, a difference in *things* brings a difference in *spaces* in its wake, even as it carries with it a difference of laws. There is, in fact, a deep analogy between Leibniz's treatment of the law-system and that of the space-system of possible worlds. And the following passage regarding laws (from a letter to Arnauld) is one I think Leibniz would apply to space as well:

> Just as there is an infinity of possible worlds, so also is there an infinity of laws, paired one for one, and every possible individual of every world includes in its notion the laws of its world.[4]

5. HOW ARE DISTINCT SPACES DISTINCT?

It is worthwhile to pose abstractly the general (and not strictly

Leibnizian) issue: Just exactly what is the cash-value difference between speaking of a plurality of *distinct* spaces as opposed to speaking of a *single* all-comprehending superspace with many distinct sectors or subspaces? And just as one is inclined to say that the reality of real physical objects resides in their localibility is one common and unified *actual* space,[5] why could one not say that the possibility of the possible lies in its localibility in one vast and all-inclusive common and unified superspace?

The answer here turns on two (interrelated) issues: Is the so-called superspace such that

(1) The various sectors bear such fundamentally *spatial* relations to one another as (for example) relative proximity and distance?

(2) The various sectors are so *connected* with one another that one can envisage some sort of "transport" within the space along an itinerary leading from each to the others?

Clearly if the answer to both of these questions is *no* – if the so-called "subspaces" are disconnected from and spatially unrelated to one another – then there is no warrant for speaking of an over-arching "superspace" at all.

The point is simply this, that a space is individuated as a single space through the mutual relatedness and connectedness of its parts, and where these elements of mutual relation and interconnection are absent, the warrant for speaking of a single space is lacking.

Now when these general considerations are brought to bear on the Leibnizian situation, it is clear that the spaces of distinct possible worlds are – or can be – so unrelated and disconnected as to remove all warrant for speaking of a single uniting space. The "Wonderland" of Lewis Carrol's Alice, the "Land of Oz" of L. Baum's stories, and the "Planet Zeta" world of the Dr Who adventures (taken as rough indications of Leibnizian worlds) are sufficiently devoid of spatial connections and relations with one another that there is no warrant for taking them as distinct sectors of a single spatial matrix. (Perhaps this claim involves a slight overstatement, since in each instance it was possible by some process – however mysterious – to transpose someone thence from *our* world.)

Different spaces cannot form parts of a unifying superspace because they must be fundamentally *disjoint* – not only in a *physical* but even in an *intellectual* sense. In the extremely interesting opuscule on space "On Existence, Dreams, and Space", Leibniz writes:

> [S]pace is that which makes that many perceptions cohere with each other at the same time . . . The idea of space is, therefore, that through which, as is recognized, we separate clearly the place, and even the world, of dreams, from ours . . . From this it follows furthermore that there can be infinitely many spaces and, hence, worlds, such that between them and ours there is to be no distance . . . Plainly as the world and space of dreams differ from ours, so too can they have other laws of motion . . . When we awake from dreams we come upon more congruences that govern bodies, but not that govern minds . . . Whoever asks whether another world, or another space, can exist is asking to this extent whether there are minds that communicate nothing to us.[6]

With Leibniz, moreover, there is a special reason why there can be no such thing as a many-world embracing superspace. We know that, for Leibniz, a substance internalizes its relations to others in the property-system that constitutes its complete individual notion. Insofar as they go beyond this property-internalization, all relations are only "things of the mind", mere *entia rationis* whose "being" is virtual and imaginary, devoid of any real existence in its own right. And this is true, in particular of spatial relations.[7] The spatial relations among substances of the same possible world – like all other relations among them – thus have at least a derivative, supervenient reality, namely that which arises through the prospect of their being realized along with their terms. But a "relationship" among the incompatible substances of different possible worlds – since they relate *incompossible* terms – can never have both feet together on the *terra infirma* of at least possible realization. It is, for Leibniz, already stretching matters to speak of spatial relations among compossibilities; to contemplate spatial relations among *incompossibles* would stretch the concept of special relatedness beyond its working limits. (As we have seen, space is "the order of *coexistence*", and incompossibles are such that – by their very nature – they cannot possibly coexist.)

6. WHY DISTINCT SPACES?

It is quite clear *why* Leibniz wants to insist on the irreconcilable distinctness of the different spaces of different possible worlds. For if those worlds could be co-located within one superspace, then it would be feasible to realize *all* possibilities by the old atomists' device of shelving each world in its appropriate spot in the all-inclusive matrix. Any prospect for an ethics of creation-choice would now be removed, and we would return to the omni-necessitarianism of Spinoza.

Leibniz develops this line of thought in an interesting essay of 1679:

But I was pulled back from this precipice by considering those possible things which neither are nor will be nor have been. For if certain possible things never exist, existing things cannot always be necessary; otherwise it would be impossible for other things to exist in their place, and whatever never exists would therefore be impossible. For it cannot be denied that many stories, especially those we call novels, may be regarded as possible, even if they do not actually take place in this particular sequence of the universe which God has chosen – unless someone imagines that there are certain poetic regions in the infinite extent of space and time where we might see wandering over the earth King Arthur of Great Britain, Amadis of Gaul, and the fabulous Dietrich von Bern invented by the Germans. A famous philosopher of our century does not seem to have been far from such an opinion, for he expressly affirms somewhere that matter successively receives all the forms of which it is capable (Descartes, *Principles of Philosophy*, Part III, Art. 47). This opinion cannot be defended, for it would obliterate all the beauty of the universe and all choice . . . [8]

What, however, of the ontological status of other spaces? They do have reality of some sort – for they "really" are the coexistence-order relations of the manifolds of possibility that they relate. But this reality is not, of course, one of actual existence. It is, at best, the purely mental existence of the orders as subjects of thought in the mind of God. "And so the reality of bodies, of space, of motion, and of time," so Leibniz writes to des Bosses in 1712, "consists in their being phenomena for God (*phaenomena Dei*) of objects of the vision of His knowledge."[9] The reality of such phenomena is merely mental – though we ought, no doubt, to hesitate just a bit in using "mere" where it is God that is at issue.

7. A SUPERSPACE AFTER ALL?

But is there not, after all, a somewhat different basis for holding Leibniz committed to a superspace theory? For spaces – all spaces – are *entia rationis* (since there is no such *substance* as a space). And (as Leibniz sees it) the entire manifold alternative possible worlds exists *in concept* in the mind of God *sub ratione possibilitatis*. Does not *God* relate the different spaces of the different possible worlds – coordinating them within one all-embracing superspace? Is not God's conception of a plurality of world-spaces tantamount to a conception of a single vast spatial matrix that embraces a plurality of parts?

Surely not. The fact that the mind of God conceives the various possible space-orders no more means that they are comprehended within one superspace than does the fact that He conceives infinitely many laws means that they are all comprehended within one superlaw or the fact that He conceives infinitely many men means that these are all parts of one superman.

A plurality of distinctly conceived spaces is something very different from the conception of a single space with a plurality of sectors. To be sure, if, for Leibniz, space were (*contra factum*) the order of what can possibly *exist* other than which can possibly *co-exist* than we would be led to a single plurality-involving superspace. But in this event the conception of space would have to play a role in Leibniz's system very different from its actual one.

The concept of a space arises from the conceptualization of spatial relations, and these relate the different items embraced *within* a common world. Spatial relations do not – and cannot – relate different possible worlds to one another spatially. And for a good reason – different Leibnizian worlds do not bear spatial relations to one another. Their "coexistence" in the mind of God is not the sort of coexistence that can give rise to a "space". As Leibniz explicitly says in the Jagodinsky passage quoted above, spaces arise out of relationships of "distance" which in turn root in the perceptions of substances, and *there are no cross-world perceptions*. The substances of distinct worlds do not have any distance from one another – not that their distance is 0, the whole concept just does not apply.

8. CROSS-WORLD SPATIAL COMPARISONS

The contention that different possible worlds have their own spaces does, however, encounter one difficulty right away – a theoretical difficulty whose bearing is general and goes outside a specifically Leibnizian context. For can we not, in fact, actually make cross-world spatial comparisons? Suppose M. Eiffel had made his tower a centimeter shorter. Clearly this diminished tower cannot be accommodated within *this* world of ours along with the actual tower. The world it inhabits is clearly *another* possible world. But surely it would still maintain various spatial relations to the thing of this world: it would still be in Paris – it would be closer to Rome than to Toronto, would it not?

In the Leibnizian setting, the answer here is not straightforward – it is yes-*and*-no.

Think back to Leibniz's treatment of comparable hypotheses, the hypothesis say that Julius Caesar had been born normally, without requiring his mother to undergo a "caesarian" section. We know how Leibniz treats this. He insists that this variant Ceasar is *not* identical with ours. The Principle of the Identity of Indiscernibles precludes strict identification. Only because of a general *resemblance* can we speak – loosely and inaccurately (popularly and without metaphysical strictness) – of this variant individual as "Julius Caesar" (i.e., as identical with the Julius Caesar of our world).

The situation with respect to space must be viewed in a strictly analogous light. If the Eiffel tower were a centimeter shorter it would NOT really be in Paris any longer – not, that is to say, in *our* actual Paris. The "Paris" in which it is located – and the "Rome" and "Toronto" to which it has spatial relations are not those of our world, but those of another world. Thus it does not, in fact, have any spatial relations to the things of our actual world.

It is surely authentic Leibnizian doctrine that spatial relationships can obtain only *within* and not *across* possible worlds.

This point becomes particularly telling if we recall the full scope and variety of Leibnizian possibilia. There are, to be sure, those possible substances which arise from hypotheses that modify *actualia* – the Adam who does not sin, the Judas who does not betray, the Caesar who does not cross the Rubicon. But not all possible substances need be variant versions of actual substances. We need not be in a position to *reidentify* a possible substance with any actual individual – not every hypothetical world is a *roman à clef* reworking of the actual one. Leibnizian possible worlds will, in general, differ very drastically from ours in their make-up – so drastically as to remove any basis for spatially relating their constituents and those of our world.

9. MUST THE SPATIAL STRUCTURE OF OTHER WORLDS BE LIKE THAT OF OURS?

But even if distinct worlds have distinct spaces, will it not, nevertheless, be the case that the spatial *structure* of other worlds will be the same – or at any rate similar – to the spatial structure of ours?

There is nothing in Leibniz's philosophy that constrains him to answer this question affirmatively. Consider again the possible world whose Eiffel Tower was built a bit shorter (say because the iron founders who made the girders worked a trifle less exactingly). Its substances are in general so similar to those of our world that its spatial structure would be virtually identical with that of ours. But this is a very specialized circumstance, one that will certainly not be realized in general.

These considerations point to an interesting question. If space were a content-indifferent container, if alternative possibilities were a matter of changing things around in one selfsame space, then clearly the truths about this space would hold in every possible world. And so geometric truths – truths about the structure of space – would be necessary. But if space is something world-relative, if different worlds would have different space-orderings, then the truths of geometry will be contingent. Just how does Leibniz view this issue of the status of geometry as necessary or contingent?

To the best of my knowledge, he does not ever address this question directly. And perhaps his principles would not quite force him to opt for contingency: Geometry, after all, deals with the *structure* of space, and it is conceivable that each member of the gamut of distinct alternative spaces would have the same geometric structures, even as very different poems could have the same metrical structure. But while this may possibly be Leibniz's position, it seems implausible that it should actually be so – given that very distinct spatial structures can be projected in imagination with relatively little effort.

In thinking of the manifold of Leibnizian possible worlds we must avoid any inclination to keep our imagination under too tight a rein. Possible worlds can differ from ours very drastically indeed. (Some, after all, might contain only a finite number of monads.) And worlds whose substances are radically different and behave in line with radically different laws of nature, might well have a spatial structure quite different from ours. For Leibniz, the truths of geometry – unlike those of arithmetic – almost certainly belong to the contingent sphere.

It seems plausible to suppose that Leibniz's project of *analysis situs* ("topology" as we nowadays call it) actually represents an attempt to devise a theory of spatial relationships that does not involve the whole range of specific commitments of a full-blown Euclidean geometry. Leibniz would surely have been neither surprised nor dismayed at the discovery of non-Euclidean geometries, and that he would have had no difficulty in assimilating such a diversity of spatial structures to his own theory of space.

10. THE IMPORTANT FACT THAT, FOR LEIBNIZ, TIME IS COORDINATE WITH SPACE

Let us now turn to time. Here we can be brief. Time, for Leibniz, is conceptually coordinate with space: one could not have space in an atemporal context, nor conversely. For space is the order of *co-existence* – that is, the order among the mutually contemporaneous states of things; while time.is the order of *succession*, that is, the order among the various different mutually coexisting states of things which, qua mutually coexisting, must of course have some sort of spatial structure.

Let me explain what is going on here by a cinematographic analogy. To be sure, Leibniz himself did not think of the matter in this naive pictorial way. But he thought of it in roughly equivalent terms – namely, in terms of mathematical analogues in the theory of real-variable functions. However (Plato notwithstanding) not all philosophers are mathematicians, and a pictorial approach may help to get the point across more effectively.

Take a motion picture film: the film reels, say, for "Gone With the Wind". And let us suppose that an immense jig-saw puzzle is created by the cutting up of this film – first into individual frames and then even more finely. The Leibnizian ordering problem is now a two-fold one, first to assemble all of the individual frames – the contemporaneity (or coexistence) slices that define its spatial order; and secondly the ordering of these contemporaneity slices into the proper sequence that defines a temporal order. For Leibniz, space and time thus stand in an inseparable co-ordination with one another in the overall ordering process that begins from the starting point of the particular states of individual substances and arrives at an all-comprehending spatio-temporal order. This coordinated symbiosis of space and time is an important aspect of Leibniz's metaphysics. For him – unlike Kant – space and time are mutually coordinate in such a way that neither is more fundamental than the other.

For present purposes, the important fact is that the factor of world-to-world variation thus comes in once again – but now with respect to time itself. For the temporal order need by no means be that of the present "Newtonian" world in which time (presumably) flows in the equable manner of a continuous parameter changing uniformly. A *discrete* time consisting of discrete discontinuous jumps, for example, is in principle, perfectly conceivable on Leibnizian principles – not, to be sure, as a condition holding in this best of possible worlds, but for one of its possible albeit suboptimal alternatives. In general, time, like space, need not be structurally uniform across possible worlds.

11. CAN A POSSIBLE WORLD LACK SPATIOTEMPORAL STRUCTURE?

We come finally to a rather delicate Leibnizian issue. Could a possible world lack having a spatiotemporal structure altogether? Could the states of its substances be in such a whirl of "blooming, buzzing confusion" that a space-time order is lacking?

Leibniz would surely argue that this cannot be – that a world cannot lack a space-time order any more than it can lack a causal order. After all, even a chaotic arrangement is some sort of "ordering" – even a random ordering is an ordering (and a very characteristic sort of ordering at that).

In the *Discourse on Metaphysics*, Leibniz formulates the issue in the following terms:

That God does nothing which is unorderly, and that it is not even possible to assume events which are not according to rule. The volitions or actions of God are commonly classified into ordinary and extraordinary acts. But it is well to understand that God does nothing without order. So whatever passes for extraordinary is so only in relation to some particular order established among creatures. For as concerns universal order, everything is in conformity with it. So true is this that not only does nothing happen in the world which is absolutely irregular but one cannot even imagine such an event. For let us assume that someone puts down a number or points on paper entirely at random, as do those who practice the ludicrous art of geomancy; I maintain that it is possible to find a geometric line whose law is constant and uniform and follows a certain rule which will pass through all these points and in the same order in which they were drawn. And if someone draws an uninterrupted curve which is now straight, now circular, and now of some other nature, it is possible to find a concept, a rule, or an equation common to all the points of the line, in accordance with which these very changes must take place. There is no face, for example, whose contour does not form part of a geometric curve and cannot be drawn in one stroke by a certain regular movement. But when the rule for this movement is very complex, the line which conforms to it passes for irregular. Thus we may say that no matter how God might have created the world, it would always have been regular and in a certain general order.[10]

To be sure, there are possible worlds so chaotic in their make-up that it would be inappropriate to characterize the relationships among the states of its substances as generating a "spatiotemporal order" *as we know it*, judging in terms of the continuities and regularities of our world. But to say this is to say little more than that the world with which we are familiar – the world we ourselves inhabit – is a very special one in the Leibnizian framework. It is, after all, the best possible world in a respect that puts prime emphasis on lawfulness and rational order.

In sum, then Leibniz holds that every possible world has a spatio-temporal structure of some sort – one that is as unique to and characteristic of it as the substances which constitute it and the laws which govern them.[11]

NOTES

[1] GP, Vol. II, pp. 450–451; Loemker, p. 604.
[2] Jagodinsky, p. 114. (Cf. note 6 below.)
[3] Letter V, §§29 and 62.
[4] GP, Vol. II, p. 40.
[5] See Anthony Quinton, "Spaces and Times," *Philosophy*, **37** (1962), 130–147. Compare also Norman Swartz, "Spatial Worlds and Temporal Worlds: Could There Be More Than One of Each", *Ratio* **17** (1975), 217–228. This discussion moots the prospect of several distinct space-time frameworks for the real. These discussions are, however, different from our present inquiry which in effect asks whether the space-time framework or frameworks for *realia* do or do not coincide with those for *possibilia.*
[6] Jagodinsky, p. 113. This opuscule was written during Leibniz's Parisian visit in 1676. It was published (with Russian translation) in I. Jagodinsky (ed.), *Leibnitia elementa philosophiae arcanae de summa rerum*, edn. (Kazan, 1913).

My access to it is through the translation and commentary prepared by H. N. Castaneda for his interesting essay "Leibniz's Meditation on April 15, 1676 about Existence, Dreams, and Space" in A. Heinekamp *et al.* (eds.) *Leibniz à Paris*, Vol. I (Wiesbaden, 1978), pp. 91–130.

7 Compare the passage of note 1, above.

8 From the essay "On Freedom"; Loemker, p. 263.

9 GP, Vol. II, p. 438.

10 *Discourse on Metaphysics*, §6; Loemker, p. 306.

11 This is a slightly revised version of a paper of the same title which appeared originally in *Rice University Studies* Vol. 63 (1977), pp. 97–106. The paper is part of an ongoing dialogue with Prof. Yvon Belaval. See his "Note sur la pluralité des espaces possibles d'après la philosophie de Leibniz", in R. Berlinger *et al.* (eds.) *Perspektiven der Philosophie*, Vol. 4 (Amsterdam, 1979), pp. 9–19.

ESSAY VII

THE CONTRIBUTIONS OF THE PARIS PERIOD (1672–76) TO LEIBNIZ'S METAPHYSICS

1. OVERVIEW OF CARDINAL THESES OF LEIBNIZ'S METAPHYSICS

This concluding essay seeks to elucidate the biographical background of the preceding discussion of Leibniz's recourse to the idea of perfection-maximization through infinitistic comparisons. In pursuing this goal, it will assess the extent to which the philosophical and mathematical work of Leibniz's Parisian period contributed to the formation of his entire metaphysical system.[1]

To determine the extent to which a part contributes to a whole we must begin by setting before our minds just what this whole actually is. In the present case, this is by no means all that simple. For there can be considerable dispute as to just what the central, salient features of Leibniz's metaphysics in fact are. Yet in the confines of a rather brief discussion one cannot but be somewhat dogmatic about this.[2] As I see it, the really major building blocks of Leibniz's monadological system are as follows:

I. *Substantival Atomism*

Thesis: The natural universe is composed of punctiform ("simple", in Leibniz's sense of this term – i.e., partless) substances differentiated by "points of view": the so-called "metaphysical points", which are indestructible and exist throughout time.

Chronology: This idea figures prominently in Leibniz's early philosophizing and was clearly developed already during the Mainz period. It is one of the fundamental ideas of his early physical tract *Theoria motus abstracti* of 1671.[3] And it is a central feature of the "geometry of indivisibles" of which Leibniz speaks in his early letter to Arnauld, and regarding which he says:

From these propositions I have reaped a great harvest, not merely in proving the laws of motion, but also in the doctrine of mind. For I demonstrated that the true locus of our mind is in a certain point or center, and from this I deduced some remarkable conclusions about the imperishability of that mind, the impossibility of ceasing from thinking, the impossibility of forgetting, and the true internal difference between motion and thought . . .[4]

II. *The Idea of a Complete Individual Concept*

A. *Concept Comprehensiveness*

Thesis: Each substance has its own characteristic nature, its proper haecceity, its own definitive individual concept. This encompasses its entire "fate" throughout the whole history of its existence – the totality of everything that happens to it – in line with the decrees of God.

Chronology: These Leibnizian doctrines are also early. His ideas on individuation go back to his youthful essay *De Principio individui*. From his earliest days as a philosopher Leibniz held that substances exist in – and throughout – time, and that their individual notion is omnitemporal in coverage and includes everything that happens to the substance. The corresponding conceptions about the necessity of a substance's descriptive make-up – that with respect to substances "whatever has happened, is happening, or will happen is best, and also necessary" is also early; for example, it is explicitly present in the letter to Magnus Wedderkopf of May 1671 (Loemker, pp. 146–147).

B. *Historical Dynamism*

Thesis: The individual notion of a substance must be conceived in dynamic terms. Not only does a substance have a developmental conatus that sweeps it along over the course of time, but its sequential unfolding is always prefigured in its earlier states. Its history is akin to the inherent dynanism of a mathematical formula with the increase of a temporal parameter, like a curve in Cartesian analytic geometry. Its temporal development accords with a characteristic relationship that embodies an infinitely varying detail of historical changes within a single all-comprising relationship.[5] Encapsulated within the individual notion of each monad is a characteristic "program" (in modern cybernetic terminology), a plan that encompasses its whole history in the manner in which differential equation synthesizes a complex course of development.

Chronology: This idea of the functional dynamics of substances is a product of the Parisian period, although it did not come to prominence until the important essay *Primae veritates* of 1680/84 (Loemker, pp. 267ff). It is worth noting that this important idea of the characterizing "program" or defining mathematical formula (or function) introduces a clearly mathematical conception into the framework of an otherwise essentially logical framework of a subject/predicate system.

III. *The Proliferation of Possibility*

Thesis: The existing (real) world is a manifold of compossible substances. This manifold of compossibles is maximal – that is, it is incapable of being augmented without engendering an incompossibility. However, the prospect of alternative compossibility manifolds remains open: some world different from this one could have been actualized. God could have created any one of infinitely many families of possibilities, seeing that an unending proliferation of possible worlds is present *sub ratione possibilatis* in the divine mind. The real existence of this actual world of ours is thus contingent; this world is in theory evitable: no impossibility is involved in the idea that another altogether different system of compossibles could have been actualized.

Chronology: This idea of the real world as one among alternatives is also an early development in Leibniz's thought. Already in May 1617 (before he had read Malebranche in Paris and had criticized Spinoza's view that all possibles exist), Leibniz wrote to Magnus Wedderkopf:

> For God wills the things which he understands to be best and most harmonious and selects them, as it were, from an infinite number of all possibilities.[6]

IV. *Creation as Maximization/Optimization*

Thesis: Creation is to be conceived of as a selective choice among alternative possible worlds for actualization. Given God's beneficence, this choice becomes a matter of maximizing goodness – of optimizing that combination of richness of phenomena and simplicity of laws which affords Leibniz's standard of ontological value.[7] Creation is the result of an act of God's will guided by the relative perfection of things (the harmonia rerum). The "boundary line problem" of what differentiates the actual from the merely possible is answered with reference to a complex essence-maximization process.

In his latish, interesting, and very important little essay *Tentamen anagogicum* (1696) Leibniz puts the matter as follows:

> The principles of mechanics themselves cannot be explained geometrically, since they depend on more sublime principles which show the wisdom of the Author in the order and perfection of his work. The most beautiful thing about this view seems to me that the principle of perfection is not limited to the general but descends also to the particulars of things and of phenomena and that in this respect it closely resembles the method of *optimal forms*, that is to say of forms *which provide a maximum or minimum*, as the case may be – a method which I have introduced into geometry in addition to the ancient method of *maximal and minimal quantities*. For in these forms or figures the *optimum* is found not only in the whole but also in each part, and it

would not even suffice in the whole without this. For example, if in the case of the curve of shortest descent between two given points, we choose any two points on this curve at will, the part of the line intercepted between them is also necessarily the line of shortest descent with regard to them. It is this way that the smallest parts of the universe are ruled in accordance with the order of greatest perfection.[8]

The workings of the Principle of Perfection resembles that of physical principles of a familiar kind. Leibniz himself makes much of such principles as the optical principles of least (or greatest, as the case may be) time and distance, and the principle of least action. The Principle of Perfection also specifies that in nature some quantity is at a maximum or a minimum. In other words it is a "minimax" or "extremal" principle, akin to the Principle of Least Action in physics.[9] It, like the others, requires techniques analogous to those of the calculus, and especially the calculus of variations. In fact, according to Leibniz's own statements, the principle was suggested to him by mathematical considerations,[10] mathematical considerations which themselves were the fruits of the Paris period, and related to the calculus, to minimax analysis and the calculus of variations (as we now call it).

Chronology: This entire essence-maximization procedure is part of the Parisian heritage of Leibniz's thinking.[11] Leibniz's conception of the deity's way of proceeding in the selection of one of the possible worlds for actualization can be represented and illustrated by the sort of infinite comparison process familiar from the calculus and the calculus of variations.

In taking as measure of perfection the combination of two essentially conflicting factors, Leibniz unquestionably drew his inspiration once again from mathematics – as he so often does. Determining the maximum or minimum of that surface-defining equation which represents a function of two real variables specifically requires those problem solving devices for which the mechanisms of the differential calculus were specifically devised. Unlike the relative mathematical naiveté of the old-line, monolithic, single-factor criterion, the Leibnizian standard of a plurality of factors in nonlinear combination demands the sort of mathematical sophistication that was second nature to him.

V. *Universal Harmony*

Thesis: Harmony, for Leibniz, is essentially a matter of the mutual attunement and accommodation of substances. Their respective states represent coordinated "points of view" that reflect a common and shared universe. And this is not only a matter of the harmonization of states or "perceptions" but also of the coordination of the changes of state with the passage of time.

Chronology: The fundamental ideas at work here are early and pre-Parisian. They are prefigured in the *Catholic Demonstrations* of 1669–70 and clearly present in the letter to Magnus Wedderkopf of May 1671 (Loemker, p. 146). The "harmony of things" is there portrayed in the role of the determinative standard for that optimization which governs the divine will in his creation-choice, and this theme recurs frequently in the Paris writings.[12] And it is clear that these ideas are readily integrated with and further developed on the basis of the more sophisticated ideas regarding optimization/maximization that were developed in the wake of the Parisian visit.

These various doctrines combine to give a reasonably complete inventory of the basic components of Leibniz's monadological system. Reasonably complete – but not quite.

2. A MISSING PIECE

The mind of any reader who has followed the preceding discussion with some inclination to over-all agreement will, no doubt, already have leaped ahead to the question: If indeed all of these major building blocks of Leibniz's metaphysics were prepared and lay ready to hand by the end of the Parisian period (1676), why did Leibniz not then begin to compile and promulgate his system? Why did his synthesizing and publicizing of this system have to wait for almost a decade, until the middle 1680s with the *Discours de mètaphysique* and the ensuing correspondence with Arnauld?

Such a question is notoriously difficult and problematic – it is always easier to explain what people do than to account for what they do not do.

Now, of course, it is always possible to see the answer to in terms of fortuitous developments. Perhaps Leibniz was just too busy to perpare an exposition of the system. Or perhaps he focussed his interests in other directions and became too preoccupied with other issues. Perhaps the occasion did not arise or the idea did not occur to him.

None of these theoretically available possibilities strike me as plausible. I do not think that any such adventitious reason of biographical happenstance was operative here. I believe there were important substantive reasons for Leibniz's holding back.

I think that Leibniz did not endeavor to compile the system and put it forward because he was dissatisfied with it – because he sensed that it was incomplete. Leibniz, I would contend, was still reluctant to promulgate his "system" because he felt that one absolutely crucial issue still remained unsolved – the issue of contingency.

Contingency was the great impediment, the stumbling block that lay in the way of a smooth synthesizing of Leibniz's ideas.

As the letter to Magnus Wedderkopf of 1671 shows, Leibniz's earliest thinking came close to an acceptance of necessitarianism.

What then is the reason for the divine intellect? The harmony of things. But what is the reason for the harmony of things? Nothing. For example, no reason can be given for the ratio of 2 to 4 being the same as that of 4 to 8, not even in the divine will. This depends on the essence itself, or the idea of things. For the essences of things are numbers, as it were, and contain the possibility of beings which God does not make as he does existence, since these possibilities or ideas of things coincide rather with God himself. Since God is the most perfect mind, however, it is impossible for him not to be affected by the most perfect harmony, and thus to be necessitated to do the best by the very ideality of things. This in no way detracts from freedom. For it is the highest freedom to be impelled to the best by a right reason. Whoever desires any other freedom is a fool. Hence it follows that whatever has happened, is happening, or will happen is best, and also necessary, but as I have said, with a necessity which takes nothing away from freedom because it takes nothing from the will and from the use of reason.[13]

As Leibniz penetrated further into the cross-currents of Cartesian philosophizing – as he entered more deeply into the thought of Malebranche and of Spinoza – he became increasingly discontent with a necessitarianism that blocked the way to genuine contingency in nature. But nevertheless, the major elements of his own doctrines all pointed in a necessitarian direction.

What alternative compossibility-systems there are is a necessary issue. Where they stand relative to each other in point of perfection is necessary. Hence, that such-and-such a one is best is necessary. Gods creating the best is necessary. Hence existence of this world is necessary. Hence it is only seeming (only phenomenal *Schein*) that there are other possibilities.

Leibniz felt a deep dissatisfaction with this condition of things. His thought was deeply committed to an authentic prospect of alternative possibilities. On this issue, Leibniz differed sharply from Spinoza and felt a keen revulsion against the position of Descartes. Leibniz emphatically did not want to have it happen – as did, in fact, happen among his interpreters as late as Russell – that he should be accused of Spinozism. This made him uncompromisingly determined that a place must be made for contingency. Accordingly, he did not feel comfortable about promulgating his teaching until he could be perfectly clear in his own mind that his system provided a secure basis for avoiding universal necessitation.[14]

That there be a valid place for contingency .... But, how – what place? I believe this to be an issue with which Leibniz grappled for a long, long time.

And for good reason. The solution to the problem was not going to be easy, seeing that most of the major doctrines pointed in a necessitarian direction. But at last Leibniz saw a way out. A combination of logical, mathematical and metaphysical ideas gave him a way out of the labyrinth. The solution of the problem of contingency within the framework of his other basic commitments – whose orientation is unquestionably necessitarian – was difficult and represents one of the major intellectual *tours de force* of Leibniz's philosophizing.

Let us examine the structure of this solution.

The starting point was an already well entrenched idea, namely that there is a determinative principle to which God subscribes in His selection of one among the possible worlds for actualization: the Principle of Perfection or of the Best.[15] In accord with this principle God selects that possible universe for which the amount of perfection is a maximum. This principle indicates that in His decision of creation God acted in the best possible way; so that the actual world is that one among the possible worlds which an infinite process of comparison showed to be the best.

But to yield contingency, this ontological principle requires a logical supplementation. We must confront the question of what Leibniz intends when he speaks of the contingent truths as analytic, but requiring an infinite process for their analysis. A given proposition concerning a contingent existence is true, and its predicate is indeed contained in its subject, if the state of affairs characterized by this inclusion is such that it involves a greater amount of perfection for the world than any other possible state; i.e., if the state of affairs asserted by the proposition is one appropriate to the best possible world.

All contingent propositions have sufficient reasons or equivalently have *a priori* proofs which establish their certainty, and which show that the connection of subject and predicate of these propositions has its foundation in their nature. But it is not the case that contingent propositions have demonstrations of necessity, for their reasons are based only on the principle of contingence or existence, i.e., on what is or seems best among the equally possible alternatives . . .[16]

An infinitistic process is thus imported into the analysis of a truth dealing with contingent existence. The analysis at issue ramifies through the infinite comparison process demanded by the Principle of Perfection.

The issue of contingency, of course, becomes particularly acute with respect to God Himself – in particular with respect to the line between his nature and his actions. Since God exists necessarily, His existence being contained in His essence, it follows that God has the highest possible degree of perfection. And so God is not only the necessary, but also, in consequence, the perfect being. But at this point we must draw attention to an equivocation in Leibniz's word "perfection", an equivocation of which he himself was perfectly well aware. There is firstly "perfection" as a measure of potentiality for existence, which we have already considered, and also "perfection" as a moral attribute, goodness. Leibniz terms the firmer metaphysical, the latter moral perfection, and he insists that these must be discriminated.[17] God is, however, perfect in both senses; as "the necessary being" He possesses the maximum amount of essence (=metaphysical perfection), and as the maxibenevolent being, His acts (the sphere of His activity being the world) are the best possible (whence moral perfection). But while God's existence, and hence His metaphysical perfection, is, as we have seen, necessary, His goodness as creator, i.e., moral perfection, is contingent and the result of free choice. "The true reason why these things rather than those exist is to be attributed to the free decrees of the divine will, the first and foremost of which is to act in all respects in the most perfect possible way, as befits the wisest of beings".[18] God's moral perfection (goodness) has a sufficient reason, and this in turn another, *et caetera ad infinitum*; but this sequence of sufficient reasons converges on God's metaphysical perfection.[19] Or, putting this another way, we can say that God's moral perfection is indeed a logical consequence of His metaphysical perfection, but a consequence which no finite deduction suffices to elicit. As Leibniz put the issue:

> If anyone asks me why God has decided to create Adam, I say, because he has decided to do the most perfect thing. If you ask me now why he has decided to do the most perfect thing, or why he wills the most perfect . . . I reply that he has willed it freely, i.e., because he willed to. So he willed because he willed to will, and so on to infinity . . .[20]

In this way, as Leibniz insists, the proposition asserting God's moral perfection is contingent; God is good by free choice, not necessitation.

Thus, it is precisely the infinite regress which Russell invokes reproachfully in his *reductio ad absurdum* of Leibniz's contention that God's goodness is contingent which establishes this contingence. Leibniz put the matter as follows:

> It is needful to distinguish between necessary or eternal truths, and factual or contingent truths. These differ pretty much as do commeasurable and incommeasurable numbers. For necessary truths can be resolved into identicals as commeasurable quantities into a common measure. But with contingent truths, as with incommeasurable numbers, the resolution proceeds *in infinitum* and is never terminated. Hence the certain and full reason of contingent-truths is known to God alone, who encompasses the infinite in a single glance (*uno intuitu*) . . .[21]

For Leibniz's theory of contingence rests on the thesis: "*Contingentiae radix est infinitum*".[22] He had mastered a lesson which philosophy was slow to learn – that infinite processes are not *ipso facto* vicious, since convergence is possible.

In his doctrine of contingence, perhaps more heavily than in any other part of his philosophy, Leibniz the philosopher is indebted to Leibniz the mathematician and to Leibniz the logician as he developed in the post Parisian period. For the infinitistic logic underlying this doctrine essentially stems from and relies upon Leibniz's mathematical investigations of the Parisian period:

> These is something which had me perplexed for a long time – how it is possible for the predicate of a proposition to be contained in (*inesse*) the subject without making the proposition necessary. But the knowledge of Geometrical matters, and especially of infinitesimal analysis, lit the lamp for me, so that I came to see that notions too can be resolvable in infinitum.[23]

And again:

> At length some new and unexpected light appeared from a direction in which my hopes were smallest – from mathematical considerations regarding the nature of the infinite. In truth there are two labyrinths in the human mind, one concerning the composition of the continuum, the other concerning the nature of freedom. And both of these spring from exactly the same source – the infinite.[24]

The adequate working out of this theory was a matter of wedding these mathematical ideas regarding infinitistic processes with logical inquiries regarding the analysis of propositions which themselves were not finally completed until after the logical work of 1679 and 1684/85. These proved to be crucial to a theory of contingency based on the idea of contingence as inherent in infinitely regressive albeit convergent analyses.

Little wonder, then, that Leibniz eventually wrote the retrospective comment "*His egregie progressus sum*" on the essay *Primae veritates* in which he first presented the finished working-out of these logical deliberations. For it was the idea of infinite analysis – a combination of logic and mathematics – which at last gave him the "thread of Ariadne" to lead out of the labyrinth of necessitation.

Once these developments gave him the necessary security, Leibniz could go back sure-footedly and write remarks like the following on his earlier sketches:

I later corrected this, for it is one thing for sins to happen infallibly, another for them to happen necessarily.[25]

These considerations accordingly combine to provide an answer to our initial question. There was, it would appear, a very good reason for Leibniz to refrain from promulgating his system in the immediate post-Paris period. For it was still seriously unfinished. The key issue of the theoretical legitimation of contingency still remained unresolved and had to await developments not achieved until several years afterwards.[26]

3. CONCLUSION

Let us review the main result of this discussion.

Two crucial contributions of the Paris period to the building-blocks of Leibniz's metaphysical system were as follows:

(1) The temporal dynamics of the complete individual notion of substances: combining change with an atemporality via the idea of a mathematical function that synthesizes the course of change.

(2) The idea of creation as maximization, and the conception of an infinite comparison process along the lines afforded by the calculus.

Their mathematical provenience is an unmistakable feature of these contributions. And this is only natural and fitting, considering Leibniz's mathematical preoccupations during the Parisian period.

The deliberations of this discussion accordingly make manifest what was perhaps obvious to start with: the deep indebtedness of Leibniz's philosophy to his work in mathematics in general, and to the mathematical labors of his Parisian period in particular. But they go beyond this truism in indicating just how large the extent of this indebtedness actually is.[27]

NOTES

[1] For the mathematical aspect, see J. E. Hofmann, *Die Entwicklungsgeschichte der Leibnizschen Mathematik während des Aufenthaltes in Paris*, Munich, 1949, as well as that author's *Leibniz's mathematische Studien in Paris* in the series *Leibniz zu seinem 300. Geburtstag*, ed. by E. Hochstetter, Berlin, 1948.
[2] For an ampler exposition see my book, *Leibniz: An Introduction to His Philosophy* (Oxford, 1979).
[3] See Loemker, p. 139.
[4] GP, Vol. I, p. 72; Loemker, p. 149.
[5] Compare the suggestion offered in Russell's Preface in the 2nd edition of *The Philosophy of Leibniz* (London, 1937).
[6] Loemker, p. 146.
[7] See the author's essay on "Leibniz and the Evaluation of Possible Worlds" given elsewhere in this volume.
[8] Loemker, p. 478.
[9] See the *Tentamen Analogicum* (GP, Vol. VII, pp. 370–379) on the minimax principles and their relation to the Principle of Perfection. The reader interested in Leibniz and the principle of least action is referred to the sixteenth note appended to Couturat's *Logique*, and to the Appendix to M. Gueroult's *Dynamique et métaphysique leibniziennes*, Paris, 1934.
[10] See below.
[11] See the Paris Notes for February 11, 1676, Loemker, p. 157; and of p. 169 (December 2, 1676).
[12] Leibniz, *Elementa philosophiae arcanae de summa rerum*, ed. by I. Jagodinsky, (Kazan, 1913) pp. 16, 28, 32, 36, etc.
[13] Loemker, p. 146.
[14] Thus in the dialogue *Confessio philosophi* (ed. Saame) Leibniz attacks the doctrine that everything is necessary (p. 58) and insists from the very outset that the acts of God are good and just (p. 32).
[15] "Principe de la perfection", "Lex melioris", "Principe du Meilleur", "Principe de la convenance".
[16] GP, Vol. IV, pp. 438–439.
[17] "And lest any should think of confounding moral perfection or goodness with metaphysical perfection or magnitude (*magnitudine*) . . . " it must be remarked that the latter is quantity of essence or reality, while the former arises when metaphysical perfection is the object of a choosing mind (GP, Vol. VII, p. 306; see also G. Grua, *G. W. Leibniz: Textes inédites*, (Paris 1948), I, p. 393). This distinction of Leibniz's has been almost universally overlooked. But if moral and metaphysical perfection are not discriminated, the

distinction between moral and metaphysical necessity collapses also, as has indeed been generally charged.

[18] GP, Vol. VII, pp. 309–310, notes. "From this, then, it becomes clear that the acts of God must be distinguished into the free and the necessitated. Thus, that God loves Himself is necessary, for it follows from the definition of God. But the God chooses the most perfect cannot be so demonstrated, for its denial implies no contradiction" (Grua, *Textes*, I, p. 288). "One can say in a certain sense that it is necessary . . . that God Himself choose what is best . . . . But this necessity is not at all at odds with contingence, it not being that necessary which I call logical, geometric, or metaphysical, whose denial is contradictory" (GP, Vol. VI, p. 284).

Until the 1680s, when his mature philosophy took form, Leibniz held a different view, as some brief tracts recently published by Grua reveal. He held that God's acts are both necessary and free. "Cum Deus necessario et tamen libere eligat perfectissimum . . . " (Grua, *Textes*, I, p. 276). God's choice it is not determined with necessity which of the alternatives is the best.

> Though it may be true that it is necessary that God chooses the best, nevertheless, it does not follow . . . that that which he chooses is necessary. because no demonstration of what is best can be given. (From *De Contingentia*, ibid., pp. 305–306.)

Thus, when H. W. B. Joseph writes, "What I should like myself to suggest by way of conclusion is, that the acts of God perhaps ought to have been declared free, but not contingent . . . " (*Lectures on the Philosophy of Leibniz*, p. 188), he suggests to Leibniz a position he did, at an early point in his career, hold. But it is not surprising that Leibniz abandoned this position. For it is difficult indeed to see how what is best could avoid being determined with necessitation when the substances are conceived *sub ratione possibilitatis*. And so we find Leibniz later flatly identifying liberty with "contingence or non-necessity" (GP, Vol. VI, p. 296).

[19] This is so since the (infinite) analysis of the contingent must ultimately lead to the necessary, i.e., to God qua metaphysically perfect (GP, Vol. VII, p. 200). It is because he holds this that Leibniz, speaking now in the language not of truths but of things, maintains, "Si nullum esset Ens necessarium, nullum foret Ens contigens" (GP, Vol. VII, p. 310).

[20] Grua, p. 302.

[21] GP, Vol. VII, p. 309.

[22] L. Couturat, *La Logique de Leibniz* (Paris 1901), p. 212 notes.

[23] L. Couturat, *Opuscules et fragments inédites de Leibniz* (Paris 1903), p. 18.

[24] L. Couturat, *La Logique de Leibniz* (Paris 1901), p. 210 notes.

[25] Loemker, p. 147, n. 2.
[26] One serious incompleteness remains. The paper deals with substance theory (substantival atomism, concept comprehensiveness, and historical dynamics) as well as world theory (proliferation of possibilities, creation as optimization, universal harmony, and contingency/infinitestic-analysis). It leaves out the theory of "middle sized" objects: physical bodies, animals, and intelligent beings. But this invokes a mixture of old and new elements that does not substantially alter the over-all proportions we have arrived at.
[27] The original version of this essay was published in A. Heinekamp *et al.* (eds.), *Leibniz à Paris: 1672–1676*, Vol. II (Wiesbaden, 1978), pp. 43–53.

APPENDIX

RESCHER ON LEIBNIZ

I should like to preface this inventory of my Leibnizian writings with a brief account of the history of my concern with the work of this fascinating and many-sided thinker, whose influence has been a recurrent *Leitmotiv* in my life. Indeed, our initial contact dates from a development of merely symbolic importance – seeing that it occurred in 1928 when I was only four or five months old – namely my parents' move to the house of my early childhood at No. 3 Leibnizstrasse in the Westphalian town of Hagen.

Though I had certainly encountered Leibniz in undergraduate history of philosophy courses, and had been intrigued by the strangeness of his system of monadology, I became seriously interested in him only after graduating from Queens College (N.Y.) in June of 1949. That summer I read Bertrand Russell's *The Philosphy of Leibniz*, and this motivated me to more serious reading and thinking about Leibniz's philosophy of mathematics and physics after I started graduate work in Princeton that fall. It was then that I discovered Louis Couturat's *La Logique de Leibniz*, and this splendid book stimulated me to work up a longish study of Leibniz's metaphysics during my first year in graduate school (the 1949–50 academic year). My interest in Leibniz dates from a time – the late 1940s – when the history of philosophy was an underdeveloped area of American scholarship, at any rate outside the classical domain. It was two figures of a past generation (Russell and Couturat) that led me to Leibniz, but it was Leibniz himself who held me there. It is perhaps fitting that I had no teacher in Leibnizian matters, but was an autodidact, relying on books alone.

As matters turned out, my Leibniz project proved to be the first draft of my 1951 doctoral dissertation on *Leibniz's Cosmology: A Reinterpretation of the Philosophy of Leibniz in the Light of His Physical Theories*. (By having stolen a march on a dissertation project in this way, it became possible for me to earn the PhD in just two years, while still only 22 years of age.)

Work on this dissertation provided the stimulus for a group of articles on Leibniz's logic and philosophy that appeared during the 1952–55 period. This work firmly implanted in my mind an ongoing interest in Leibniz's ideas and projects.

During my years at Lehigh University (1957–61), I taught Leibniz from time to time in courses in the history of modern philosophy, but otherwise my Leibnizian interests lay fallow during the 1956–66 decade. Upon coming to the University of Pittsburgh as professor in 1961, however, a graduate seminar on Leibniz became a regular feature of my teaching repertoire. This kept my interest alive, and in 1967 issued in my exposition of *The Philosophy of Leibniz*, an attempt to meet the need for a well-rounded introduction for advanced students.

Now too I began to be involved in various organizational efforts. In November of 1967 the Leibniz scholars of the world gathered in Hanover, Germany to commemorate the 250th anniversary of his death. Dr Wilhelm Totok, the able and enterprising head of the Nether-Saxon State Library, was the leading spirit behind this celebration, and on its occasion he took the decisive steps towards launching the International Leibniz Society. It was founded at a meeting in the Stadthalle at which I was present, and I was chosen a member of its Council (Beirat), an office I continue to hold to the present day. I also became a member of the editorial board of the official journal of this society, *Studia Leibnitiana*. (Indeed, it was my plea that a Latin rather than German title be adopted – for reasons of internationalism and in homage to Leibniz's boundary-transcending spirit – that led to the selection of this title in the place of *Leibniz-Studien.*) Since that time I have participated regularly in sessions of this organization held in various European cities.

When the idea of an American Leibniz Society first came to be mooted in 1976, I was also involved. I helped to organize the society, became a member of its Executive Committee, and handled the negotiations that led to its affiliation with the International Leibniz Society.

During the 1970s, I maintained a steady interest in Leibniz, writing, on the average, one paper every two years. And at the end of the decade, in 1979 I published *Leibniz: An Introduction to His Philosophy*, a revised and expanded version of my earlier book, taking account of the active literature that had sprung up during the intervening time. Throughout this period I was a regular participant at various national and international Leibniz congresses and conferences, and became an active member of that small international fraternity of members who interest themselves in the wide-ranging thought and multifarious doings of this great man. Gradually I have become a dedicated Leibnizian, though, to be sure, I am not a Leibniz-scholar in that mode of total commitment which typifies the European philosophy-historians of the old school.

I feel that over the years I have gotten to know Leibniz well and not only as a thinker but as a person. At any rate, I have learned, bit by bit, a great deal about his life and times and about the people amongst whom he labored. And repeated visits to Hanover have made me feel very much at home among the few remnants of that much smaller town with which Leibniz was so thoroughly familiar.

In closing, I should like to add a few words about Leibniz's influence on my own philosophical work. I do not view myself as an adherent of his teaching or doctrine, but rather of his mode of philosophizing. Leibniz is to my mind the master of us all in the use of the formal resources of symbolic thought in the interest of the clarification and resolution of philosophical issues – a role-model for the way in which one wants to do one's philosophical work. Accordingly, many of my books exhibit the tendency to use some formal notational or symbolic mathematical or logical or diagrammatic device for the elucidation of philosophical issues. In any case, I hope and believe that readers will sometimes find the impetus to clarity and the wide sympathies so typical of Leibniz astir in the pages of my own philosophical books.

Moreover, I have always felt a certain spiritual affinity with Leibniz, in that we have both faced a broadly common situation in a broadly common way. We both came to philosophy after a generation of iconoclasm – represented by Cartesianism in his case, and logical positivism in mine. In each instance, there had been a prior phase of annihilation of traditions and search for a fresh start, for building up everything anew on a novel foundation erected on the ruins of the older structures. This created a common mission for our philosophical generations: to build a bridge across the rubble left by our demolitionist predecessors – a bridge able to reach the far bank of the rich philosophical heritage of the past that lay beyond.

BIBLIOGRAPHY

1. "Contingence in the Philosophy of Leibniz", *The Philosophical Review* **61** (1952), 26–39.
2. "Leibniz's Interpretation of His Logical Calculi", *The Journal of Symbolic Logic* **18** (1954), 1–13.
3. Review of the Lucas-Grint edition of Leibniz's *Discourse on Metaphysics, The Philosophical Review* **63** (1954), 441–444.
4. Review of R. M. Yost, Jr., *Leibniz and Philosophical Analysis, The Philosophical Review* **64** (1955), 492–494.
5. "Monads and Matter: A Note on Leibniz's Metaphysics", *The Modern Schoolman* **33** (1955), 172–175.

6. "Leibniz's Conception of Quantity, Number, and Infinity", *The Philosophical Review* **64** (1955), 108–114.
7. "Leibniz and the Quakers", *Bulletin of Friends Historical Association* **44** (1955), 100–107.
8. Review of Pierre Burgelin, *Commentaire du 'Discours de Metaphysique' de Leibniz*, *Erasmus* **13** (1960), 385–388.
9. "Logische Schwierigkeiten der Leibnizschen Metaphysik", *Atken des Internationalen Leibniz Kongresses*, Hannover, 1966, Vol. I, *Metaphysik Monadenlehre* (Weisbaden, 1968), pp. 253–265. *Studia Leibnitiana: Supplementa*, Vol. I. Tr. in N. Rescher, *Essays in Philosophical Analysis* (Pittsburgh, 1969), pp. 159–170; reprinted in *The Philosophy of Leibniz and the Modern World*, ed. by I. Leclerc (Nashville, 1973), pp. 176–188.
10. *The Philosophy of Leibniz*, Englewood Cliffs: Prentice Hall, 1967.
11. Review of A. T. Tymieniecka, *Leibniz' Cosmological Synthesis, The Philosophical Review* **76** (1967), 244–245.
12. "Leibniz and the Evaluation of Possible Worlds", in N. Rescher, *Studies in Modality*, Oxford: Basil Blackwell, 1974, pp. 57–70.
13. "Leibniz, Gottfried Wilhelm", in *The Encyclopedia Americana* (New York, 1975).
14. "Leibniz and the Plurality of Space-Time Frameworks", *Rice University Studies* **63** (1977), 97–106.
15. "The Contributions of Leibniz's Paris Period to the Development of His Philosophy", *Leibniz à Paris* (1672–1676), Vol. II, *La Philosophie de Leibniz* (Wiesbaden, 1978), pp. 43–53. *Studia Leibnitiana: Supplementa*, Vol. XVIII.
16. "The Epistemology of Inductive Reasoning in Leibniz", *Theoria Cum Praxi: Akten des III. Internationalen Leibniz-Kongress*, Vol. III (Wiesbaden, 1980). *Studia Leibnitiana: Supplementa*, Vol. XXI.
17. *Leibniz: An Introduction to His Philosophy*, Oxford: Basil Blackwell, 1979. Co-published in the U.S.A. by Rowman-Littlefield.

INDEX OF NAMES

SUBJECT INDEX

THE UNIVERSITY OF WESTERN ONTARIO SERIES IN PHILOSOPHY OF SCIENCE

A Series of Books in Philosophy of Science, Methodology, Epistemology, Logic, History of Science, and Related Fields

1. J. Leach, R. Butts, and G. Pearce (eds.), *Science, Decision and Value.* Proceedings of the Fifth University of Western Ontario Philosophy Colloquium, 1969. 1973, vii + 213 pp.
2. C. A. Hooker (ed.), *Contemporary Research in the Foundations and Philosophy of Quantum Theory.* Proceedings of a Conference held at the University of Western Ontario, London, Canada, 1973. xx + 385 pp.
3. J. Bub, *The Interpretation of Quantum Mechanics.* 1974, ix + 155 pp.
4. D. Hockney, W. Harper, and B. Freed (eds.), *Contemporary Research in Philosophical Logic and Linguistic Semantics.* Proceedings of a Conference held at the University of Western Ontario, London, Canada. 1975, vii + 332 pp.
5. C. A. Hooker (ed.), *The Logico-Algebraic Approach to Quantum Mechanics.* 1975, xv + 607 pp.
6. W. L. Harper and C. A. Hooker (eds.), *Foundations of Probability Theory, Statistical Inference, and Statistical Theories of Science*, 3 Volumes. Vol. I: *Foundations and Philosophy of Epistemic Applications of Probability Theory.* 1976, xi + 308 pp. Vol. II: *Foundations and Philosophy of Statistical Inference.* 1976, xi + 455 pp. Vol. III: *Foundations and Philosophy of Statistical Theories in the Physical Sciences.* 1976, xii + 241 pp.

8. J. M. Nicholas (ed.), *Images, Perception, and Knowledge.* Papers deriving from and related to the Philosophy of Science Workshop at Ontario, Canada, May 1974. 1977, ix + 309 pp.
9. R. E. Butts and J. Hintikka (eds.), *Logic, Foundations of Mathematics, and Computability Theory.* Part One of the Proceedings of the Fifth International Congress of Logic, Methodology and Philosophy of Science, London, Ontario, Canada, 1975. 1977, x + 406 pp.
10. R. E. Butts and J. Hintikka (eds.), *Foundational Problems in the Special Sciences.* Part Two of the Proceedings of the Fifth International Congress of Logic, Methodology and Philosophy of Science, London, Ontario, Canada, 1975. 1977, x + 427 pp.
11. R. E. Butts and J. Hintikka (eds.), *Basic Problems in Methodology and Linguistics.* Part Three of the Proceedings of the Fifth International Congress of Logic, Methodology and Philosophy of Science, London, Ontario, Canada, 1975. 1977, x + 321 pp.
12. R. E. Butts and J. Hintikka (eds.), *Historical and Philosophical Dimensions of Logic, Methodology and Philosophy of Science.* Part Four of the Proceedings of the Fifth International Congress of Logic, Methodology and Philosophy of Science, London, Ontario, Canada, 1975. 1977, x + 336 pp.
13. C. A. Hooker (ed.), *Foundations and Applications of Decision Theory*, 2 volumes. Vol. I: *Theoretical Foundations.* 1978, xxiii+442 pp. Vol. II: *Epistemic and Social Applications.* 1978, xxiii+206 pp.
14. R. E. Butts and J. C. Pitt (eds.), *New Perspectives on Galileo.* Papers deriving from and related to a workshop on Galileo held at Virginia Polytechnic Institute and State University, 1975. 1978, xvi + 262 pp.
15. W. L. Harper, R. Stalnaker, and G. Pearce (eds.), *Ifs. Conditionals, Belief, Decision, Chance, and Time.* 1980, ix + 345 pp.
16. J. C. Pitt (ed.), *Philosophy in Economics.* Papers deriving from and related to a workshop on Testability and Explanation in Economics held at Virginia Poly-Technic Institute and State University, 1979. 1981.